Lecture Notes in Mathematics

Edited by A. Dold and B. Ec

671

R. T. Smythe
John C. Wierman

First-Passage Percolation on the Square Lattice

Springer-Verlag
Berlin Heidelberg New York 1978

Authors
R. T. Smythe
Department of Mathematics
University of Oregon
Eugene, OR 91403/USA

John C. Wierman
School of Mathematics
127 Vincent Hall
University of Minnesota
Minnesota, MN 55455/USA

Library of Congress Cataloging in Publication Data
Main entry under title:

First-passage percolation on the square lattice.

(Lecture notes in mathematics ; 671)
Bibliography: p.
Includes index.
1. Limit theorems (Probability theory) 2. Renewal theory. 3. Matrices. I. Smythe, Robert T., 1941- II. Wierman, John C., 1949- III. Title: Percolation on the square lattice. IV. Series: Lecture notes in mathematics (Berlin) ; 671.
QA3.L28 no. 671 [QA273.67] 510'.8s [519.2] 78-13679

AMS Subject Classifications (1970): 60F15, 60K05, 60K99, 94A20

ISBN 3-540-08928-4 Springer-Verlag Berlin Heidelberg New York
ISBN 0-387-08928-4 Springer-Verlag New York Heidelberg Berlin

2141/3140-543210

Preface

The mathematical study of percolation processes has now been going on for some twenty years. Most of this work has concentrated on the case in which the bonds (or sites) have only two states, open or closed; we have called this the Bernoulli percolation model. First-passage percolation theory, initiated by Hammersley and Welsh in 1965, may be viewed as a generalization of Bernoulli percolation, permitting the passage times along bonds to have any non-negative distribution with a finite mean. There are in the literature a number of expositions of various aspects of Bernoulli percolation, including those of Shante and Kirkpatrick (1971), Essam (1972), Welsh (1977), and Seymour and Welsh (1977). By contrast, much of the work on first-passage percolation is quite recent and no reasonably comprehensive account has been published.

The present work is an attempt to fill this gap. Because our principal interest is in first-passage percolation, no attempt has been made to provide an exhaustive treatment of Bernoulli percolation (we have, however, presented most of the known rigorous results in Chapter III). Again, since our emphasis is primarily mathematical, we have not attempted to summarize the large literature on Monte Carlo studies of percolation, but have contented ourselves with a few references in section 3.10. Finally, we have restricted consideration to the square lattice, largely for reasons of mathematical tractability. It will be clear, however, that a number of the results or techniques presented can be applied to other lattices as well (for example, the analysis of section 3.9 will extend to other regular planar lattices, and the point-to-point first-passage processes for the cubic lattice are subadditive).

Our aim is twofold: to make available to potential users of percolation models the latest mathematical results in the area, and to suggest to the mathematical community that some interesting (and possibly useful) mathematical problems concerning percolation remain unsolved. With the first audience in mind, we have tried to make the exposition as self-contained as possible and to keep the mathematics simple. A working knowledge of probability theory at the level of Chung (1974) should be sufficient for most purposes.

Most of the mathematical results presented here have appeared (or are about to appear) in the mathematics or physics literature; we have attempted to give credit where credit is due, and we offer apologies to those who have been slighted. A considerable amount of new material appears in Chapter VIII (particularly §§ 8.2, 8.4, 8.5, 8.7, and 8.8) and in Chapter III (§§ 3.3, 3.6, 3.8, and 3.9); lesser amounts will be found in Chapters VI (§6.2) and IX (§9.1).

We wish to thank Professors P. D. Seymour, W. Reh, and D. J. A. Welsh for making some of their work available to us in advance of publication; the National Science Foundation for support under grants MCS - 7701845 (R.T.S.) and MCS - 7405786 (J.C.W.); the University of Minnesota Graduate School for support under Research Grant 494-0350-4909-

(J.C.W.); and finally our typist, Linda McClure, for her patience and good humor in what has sometimes seemed an endless task.

R. T. Smythe John C. Wierman

Eugene, Oregon Minneapolis, Minn.

Table of Contents

Chapter I - Introduction

1.1 The percolation model

In its most general form, a percolation process is a mathematical model of the random spread of a fluid through a medium, where the terms "fluid" and "medium" are to be broadly interpreted. The physical processes to be modelled might include the penetration of a porous solid by a liquid, or the spread of an infectious disease through a community of organisms. In contrast to the well-known diffusion processes, in a percolation process the random mechanism governing the spread of the fluid is ascribed to the medium, rather than to the fluid itself. Many examples of these processes are given by Broadbent and Hammersley (1957), Frisch and Hammersley (1963), and Shante and Kirkpatrick (1971); since our concern is primarily with the mathematical properties of these processes, we shall not attempt to describe the underlying physical processes in any detail.

The mathematical study of percolation processes was initiated by Broadbent and Hammersley (1957). To render the problem mathematically tractable, the medium is generally taken to have a regular lattice structure,represented by a connected graph with a finite or countable set of sites $\{x_i\}_{i=1}^{\infty}$ and bonds ℓ_{ij} joining x_i to x_j. (For given i,j, the bond ℓ_{ij} may or may not exist.) The fluid may then be thought of as being introduced at one or more "source sites" x_i and spreading to other sites along the bonds ℓ_{ij}.

The bonds may be oriented - permitting passage of the fluid in one direction only - or unoriented, allowing passage both ways.

In the original formulation two different random mechanisms were considered. In the site percolation problem, each site has (independently of all other sites) probability p of being open and probability $q = 1 - p$ of being blocked, or closed. Fluid is introduced at the source sites, and flows along the bonds (subject to orientation) to open sites only. In the bond percolation problem, each bond has (independently of all others) probability p of being open and $1 - p$ of being closed; fluid flows from the source atoms along open bonds only (again subject to orientation). In both problems, the general idea is to make (probability) statements about the set of sites reached by the fluid.

The two problems are not unrelated; in fact there is a well-known transformation ([10]) which will convert the bond problem on a planar lattice to a site problem on an associated two-dimensional lattice called the covering lattice. When all bonds of the original lattice L are unoriented, the covering lattice L^* has its sites in one-to-one correspondence with the bonds of L. The bonds of L^* are all unoriented, and are defined by the rule that two sites β and β' of L^* are to be connected by a bond of L^* if and only if their corresponding bonds α and α' of L have a common terminal site of L. However, not every site problem can be converted to a bond problem; in this sense the site problem is the more general.

The bond percolation problem described above will be hereinafter referred to as Bernoulli percolation to distinguish it from the more general case we shall now introduce.

Let L denote the simple quadratic lattice (or square lattice, as we shall call it). The sites of this lattice are the integer pairs (x,y), and the bonds are the unit length horizontal and vertical arcs joining (x,y) with $(x \pm 1, y)$ and $(x, y \pm 1)$ respectively. All bonds are unoriented. With the set of bonds $\{\ell_i\}_{i=1}^{\infty}$ we associate a sequence $\{U_i\}_{i=1}^{\infty}$ of independent, identically distributed random variables with a finite mean $\bar{u}$; the U_i will generally be taken to be non-negative, but on occasion this restriction will be relaxed. When the U_i are non-negative we may think of them as representing the time needed for a particle to move along the length of ℓ_i from one endpoint to the other. Given two sites x_i and x_j, we can then consider the total travel time from x_i to x_j along any connected path joining the two sites. The problem is then to make probability statements about the path with the shortest travel time, denoted t_{ij}.

This generalized bond percolation problem on the square lattice was introduced by Hammersley and Welsh (1965); they called the time t_{ij} the first-passage time from x_i to x_j, and proved many results about it and other related times. The central concern of this monograph is with first-passage percolation on the square lattice; when the distribution of the U_i is taken to be Bernoulli, we are thus dealing with the bond percolation problem on L, rather than the site percolation problem.

The rest of this chapter is devoted to graph - theoretic preliminaries and the general formulation of first-passage percolation. Chapter II collects some preliminary results which are needed later on. In Chapter III Bernoulli percolation is studied, both for its own sake and for use later. The first-passage processes are introduced, and their integrability properties studied, in Chapter IV; in Chapter V ergodic theorems are proved for these processes, and Chapter VI is devoted to generalized renewal theory. Chapter VII discusses some properties of the ergodic limit of the first-passage processes. In Chapter VIII the length and height of optimal routes are discussed. Chapter IX treats briefly some other percolation models on the square lattice, and we close in Chapter X with a discussion of conjectures and open problems.

1.2 General formulation of first-passage percolation on L

We begin with some basic terminology of graph theory. No attempt will be made, here or in Chapter II, to present graph-theoretic results in great generality; only what is needed for percolation processes is discussed.

If G is a graph, finite or infinite, let $V = V(G)$ denote its set of vertices (sites) and $E = E(G)$ its set of edges or bonds. A path on the graph G is an alternating sequence of vertices and edges, $v_0, e_1, v_1, e_2, \ldots, v_{n-1}, e_n, v_n$, beginning and ending with vertices, in which each edge e_i connects the two vertices v_{i-1} and v_i. A path is said to be self-avoiding (or non-self-intersecting) if all the vertices, and thus all the edges, are distinct. A circuit is a path for which $v_o = v_n$ and which has n distinct vertices.

An important property of certain graphs is connectedness. A graph is connected if every pair of vertices is joined by a path of bonds in the graph. If a graph is not connected, it has more than one maximal connected component, and is said to be disconnected. We define a cut set of a connected graph as a collection of bonds whose removal from the graph results in a disconnected graph, but the removal of any proper subset of which leaves the resulting graph connected.

In the Bernoulli percolation model we define an open (closed) cluster to be a maximal connected subgraph which has all its bonds open (closed).

Now take $(\Omega,\mathcal{B},P)$ to be a fixed probability space. Let L be the square lattice of §1.1 and $\{\ell_i\}_{i=1}^{\infty}$ the set of bonds. Let $\{U_i\}_{i=1}^{\infty}$ be an independent, identically distributed sequence of random variables on $(\Omega,\mathcal{B},P)$. The random variable U_i will be called the <u>time coordinate</u> of ℓ_i. The sequence $\{U_i\}_{i=1}^{\infty}$ defines a configuration of time coordinates on L which may be represented as an infinite vector $\omega = (U_1(\omega), U_2(\omega), \ldots.)$ called the <u>time state</u> of L under ω.

Given a path r containing bonds $\ell_{i_1},\ldots\ell_{i_n}$, the <u>travel time</u> of r under ω is defined as

(1.1) $$t(r,\omega) = U_{i_1}(\omega) + \ldots + U_{i_n}(\omega)$$

If R is any non-empty set of paths on L, the <u>first-passage time</u> of R under ω is defined by

(1.2) $$t_R(\omega) = \inf\{t(r,\omega) : r \in R\}$$

If there exists $r_0 \in R$ such that $t(r_0,\omega) = t_R(\omega)$, we say that <u>a route exists</u> for $t_R(\omega)$ and r_0 is this route. (Note that r_0 need not be unique).

Since the set of finite subsets of a countable set is countable, any set of paths R is countable, and t_R as defined in (1.2) is thus a random variable.

The next lemma is obvious:

<u>Lemma 1.1</u> If $R_1 \subseteq R_2$, then $t_{R_1}(\omega) \geq t_{R_2}(\omega)$.

If two paths $r_1 = (v_o^1, e_1^1, v_1^1, e_2^1, \ldots, v_{n_1}^1)$ and $r_2 = (v_o^2, e_1^2, v_1^2, e_2^2, \ldots, v_{n_2}^2)$ have the property that $v_{n_1}^1 = v_o^2$, the path $(v_o^1, e_1^1, \ldots, v_{n_1}^1, e_1^2, \ldots v_{n_2}^2)$ is denoted $r_1 * r_2$. Two sets of paths R_1 and R_2 will be called <u>connected</u> with connection $R_3 = R_1 * R_2$ if, given any $r_1 \in R_1$ and $r_2 \in R_2$, then $r_1 * r_2 \in R_3$, and if every $r_3 \in R_3$ equals $r_1 * r_2$ for some $r_1 \in R_1$, $r_2 \in R_2$.

The next lemma is almost as obvious as the first:

<u>Lemma 1.2</u> If R_1 and R_2 are connected sets of paths, then $t_{R_1}(\omega) + t_{R_2}(\omega) = t_{R_1 * R_2}(\omega)$.

The term "percolation process on L", unless otherwise indicated, will always refer to the model described above.

Chapter II - Preliminaries

In this chapter we will assemble some of the basic tools for the study of percolation processes. In a few cases where a result is well-known, proofs are either relegated to the Appendices or simply referenced. As mentioned in the preface, knowledge of basic probability theory is taken for granted throughout.

2.1 Whitney's theorem and percolation on the sponge.

For any planar graph G, there is a corresponding *Whitney dual graph* G^* constructed as follows: Place a site of G^* in each region enclosed by bonds of G (including the exterior region if G is a finite graph). If two regions enclosed by G have a bond e as a common edge, the sites of G^* in these regions are joined by a bond e^* in G^* which crosses e. Thus each bond in G is crossed by exactly one bond in G^*.

According to this definition, the Whitney dual of a dual graph G^* is the original graph G.

For the square lattice L, we may take the vertices of the Whitney dual L^* to be the points $(i+\frac{1}{2}, j+\frac{1}{2})$ for i and j integers, with the crossing bonds being horizontal and vertical line segments of length one perpendicular to the bonds of L. We thus arrive at an important fact for our application of graph theory to percolation problems on the square lattice: The Whitney dual of the square lattice is again a square lattice, or, in other words, the square lattice is self-dual.

There is an interesting relation between G and G^* which is critical for our purposes.

> Theorem 2.1 If G and G^* are Whitney dual graphs, then any circuit in one graph corresponds to a cut set of bonds in the other, and conversely, any cut set in one corresponds to a circuit in its dual.

This theorem, which is well-known, has a highly intuitive content; the correspondence clearly is that the set of bonds in the dual G^* which cross bonds in a circuit of G forms a cut set, while crossing bonds of a cut set in G form a circuit in G^*. We omit the details of the proof, which can be found in Whitney's papers (1932, 1933).

In the first-passage percolation model, a bond of a dual graph is always assigned the time coordinate of the bond it crosses in the original graph. In particular, in the Bernoulli percolation model, a bond in a dual graph is open if and only if it crosses an open bond.

Whitney's theorem allows us to prove a basic result about the sponge percolation model, which was developed by Seymour and Welsh (1977).

The $m \times n$ sponge $T(m,n)$ consists of all vertices and bonds of the square lattice which are contained in the region where

$1 \leq x \leq n$ and $1 \leq y \leq m$. Each bond in $T(m,n)$ is open with probability p and closed with probability $1-p$, independent of all other bonds.

Each of the m vertices $(1,y)$, for $1 \leq y \leq m$, on the left edge of the sponge is considered to be a source site for fluid, which then flows along open bonds in the sponge. The natural problem that arises is the determination of the probability that the fluid reaches the right edge of the sponge.

Given any finite rectangular portion B of the lattice L, where bonds are (independently) open with probability p, the notation $\overset{B}{\rightarrow}$ will be used for the event that there is a path along open-bonds from the left edge of B to the right edge. The notation $\updownarrow B$ is then defined in the obvious way.

For the sponge $T(m,n)$, we are therefore interested in $P(\overset{T(m,n)}{\longrightarrow}) = S_p(m,n)$. We have, trivially:

$$(2.1) \qquad S_p(m,n+1) \leq S_p(m,n)$$

$$(2.2) \qquad S_p(m,n) \leq S_p(m+1,n).$$

Define the dual sponge $T^*(m,n)$ as the graph consisting of all vertices and bonds of the dual lattice L^* contained in the region $3/2 \leq x \leq n - \frac{1}{2}$ and $\frac{1}{2} \leq y \leq m + \frac{1}{2}$. (Note that $T^*(m,n)$ is not precisely the Whitney dual graph of $T(m,n)$, and that $T^*(m,n)$ is isomorphic to $T(n-1,m+1)$.) A bond in $T^*(m,n)$ is open if and only if the bond it crosses in $T(m,n)$ is open.

The fundamental result for the sponge, which we shall use in other percolation problems, was noted by Seymour and Welsh:

<u>Theorem 2.2</u> Suppose $0 \leq p \leq 1$ and $q = 1 - p$. For all $m \geq 1$ and $n \geq 2$,

$$S_p(m,n) + S_q(n-1,m+1) = 1.$$

<u>Proof</u>. We modify the sponge $T(m,n)$ to obtain a new graph $G(m,n)$ by the following construction. Remove all the bonds on the lines $x = 1$ and $x = n$. Identify all the vertices on the line $x = 1$ as a new vertex v_1, and all vertices on the line $x = n$ as a new vertex v_2. Finally, add a new bond e joining v_1 and v_2.

To associate the graph $G(m,n)$ with the percolation model on the sponge $T(m,n)$, we assign e to be closed, with every other bond of $G(m,n)$ having the same assignment as its associated bond in $T(m,n)$. Notice that the event that an open path exists which crosses the sponge from $x = 1$ to $x = n$ corresponds exactly to the event that v_1 and v_2 are in the same cluster of open bonds in $G(m,n)$.

Suppose that v_1 and v_2 are in distinct open clusters in $G(m,n)$ (the cluster could of course consist of a single vertex). Then the removal of all closed bonds from $G(m,n)$ disconnects the graph, so the set of closed bonds contains a cut set which leaves v_1 and v_2 in distinct components. Obviously, this cut set must contain the bond e.

By Theorem 2.1, this cut set corresponds to a circuit of closed bonds in the Whitney dual graph $G^*(m,n)$. The closed circuit must contain e^*, the bond which crosses e, so it must contain a closed path across the dual sponge $T^*(m,n)$ from top to bottom.

Since v_1 and v_2 are either in the same open cluster or in distinct open clusters, the probabilities of these two disjoint events sum to one. But $T^*(m,n)$ is isomorphic to $T(n-1,m+1)$ which has each bond closed with probability $q=1-p$. Thus

$$S_p(m,n)+S_q(n-1,m+1)=1.$$

From the proof of Theorem 2.2 we deduce the following corollary:

> Corollary 2.3 For the $m\times n$ sponge $T(m,n)$, either there exists a path of open bonds which crosses $T(m,n)$ from left to right, or there exists a path of closed bonds which crosses the dual $T^*(m,n)$ from top to bottom.

2.2 The FKG inequality

Harris (1960) proved a very useful correlation inequality for a percolation model on a finite graph. This result was generalized by Fortuin, Kasteleyn, and Ginibre (1971) to a correlation inequality for monotone functions on any finite distributive lattice. This section will introduce the needed terminology and state these results;

a proof of the FKG inequality is given in Appendix A.

If two random variables X and Y satisfy the condition $E(XY) \geq E(X)E(Y)$ we will say they are covariant. The definition extends to events by defining events A and B to be covariant if and only if their indicator functions are covariant. Thus if $P(B) > 0$, the events A and B are covariant if and only if $P(A|B) \geq P(A)$. A collection of random variables $\{X_1,\ldots,X_n\}$ is called covariant if for every subset $I \subseteq \{1,2,\ldots,n\}$,

$$E\left(\prod_{i \in I} X_i\right) \geq \prod_{i \in I} E(X_i).$$

The FKG inequality will be stated for lattices. A lattice is a partially ordered set in which any pair of elements x and y have both a least upper bound $x \vee y$ and a greatest lower bound $x \wedge y$. The lattice is called distributive if the $\wedge$ and $\vee$ each distribute over the other.

If Γ is a distributive lattice, a function $f:\Gamma \to R^1$ is called increasing (decreasing) if $f(x) \leq f(y)$ $(f(x) \geq f(y))$ whenever $x \leq y$ $(x \geq y)$. If Γ is a finite lattice and μ a non-negative function on Γ, we define the μ-average of a function $f:\Gamma \to R^1$ as

$$E_\mu(f) \equiv \left[\sum_{x \in \Gamma} \mu(x) f(x) / \sum_{x \in \Gamma} \mu(x)\right]$$

Here then is the statement of the FKG inequality

Theorem 2.4 Let Γ be a finite distributive lattice and let μ be a non-negative function on Γ satisfying

(2.3) $\mu(x)\mu(y) \leq \mu(x \wedge y)\mu(x \vee y)$ for all $x,y \in \Gamma$.

If f and g are both increasing (or decreasing) functions on Γ, they are covariant, i.e.

(2.4) $E_\mu(fg) \geq E_\mu(f)E_\mu(g)$.

Our first application of the FKG inequality is the proof of a special case of Harris' Lemma 4.1 on combinations of open bonds.

Let G be a finite graph with edge set E. For each subset A of edges define $\mu(A) = p^{|A|}(1-p)^{|E \setminus A|}$, where $|A|$ denotes the cardinality of the set A, and $0 \leq p \leq 1$. Let Γ be the distributive lattice of subsets of E, with the partial ordering of set inclusion. Then μ satisfies the convexity condition (2.3).

The lattice Γ may be identified with the bond percolation model on G, when p is the probability of a bond being open; $\mu(A)$ is the probability that A is the subset of open bonds.

For vertices v_1 and v_2 in G, we let $\{v_1 \overset{G}{\rightsquigarrow} v_2\}$ denote the event that there is a path of open bonds in G connecting v_1 with v_2 (when considering the infinite lattice L, we write simply $\{v_1 \rightsquigarrow v_2\}$ to denote the event that there is a path of open bonds connecting v_1 with v_2). For any vertices v_1, v_2,

v_3, and v_4 of G, the indicator functions of the events $\{v_1 \overset{G}{\rightsquigarrow} v_2\}$ and $\{v_3 \overset{G}{\rightsquigarrow} v_4\}$ are increasing functions on Γ. An application of Theorem 2.4 thus proves the following result:

Lemma 2.6 For any vertices v_1, v_2, v_3, v_4 of a finite graph G with the bond percolation model,
$P[v_1 \overset{G}{\rightsquigarrow} v_2, v_3 \overset{G}{\rightsquigarrow} v_4] \geq P[v_1 \overset{G}{\rightsquigarrow} v_2] P[v_3 \overset{G}{\rightsquigarrow} v_4]$.

The lemma may be easily extended to deal with more than two pairs of vertices. We will typically apply Lemma 2.5 to a finite subgraph of the square lattice.

The lemma below is noted for reference in Chapter III:

Lemma 2.5 Let A_1 and A_2 be covariant events with $P(A_1) = P(A_2)$. Then
$$P(A_1) \geq 1 - [1 - P(A_1 \cup A_2)]^{\frac{1}{2}}.$$

Proof. Write $P(A_1 \cup A_2) = P(A_1) + P(A_2 | A_1^c) P(A_1^c)$. Since A_1 and A_2 are covariant,
$$P(A_2 | A_1{}^c) \leq P(A_2).$$
Thus $P(A_1 \cup A_2) \leq P(A_1) + P(A_2)[1 - P(A_1)] = 2P(A_1) - P^2(A_1)$ since $P(A_1) = P(A_2)$. So
$$1 - P(A_1 \cup A_2) \geq [1 - P(A_1)]^2$$
which yields the conclusion.

There are by now many extensions of the FKG inequality to infinite lattices; in particular, a theorem of Fortuin (1972) extends the covariance inequality to apply to monotone functions in a percolation model on a countable graph G.

2.3 Subadditive processes

The theory of subadditive processes is indispensable for the study of first-passage percolation; in fact these processes were introduced by Hammersley and Welsh (1965) for just this purpose. The definition we give below of these processes follows Kingman (1968) and is slightly more restrictive than the original definition in [27].

Let $(\Omega,\mathcal{B},P)$ be a probability space, N the set of non-negative integers. A collection of random variables $\{X_{mn}: m,n \in N, m<n\}$ defined on $(\Omega,\mathcal{B},P)$ is called a subadditive process if it satisfies (2.5) - (2.7) below:

(2.5) If $m<n<p$, $X_{mp} \leq X_{mn} + X_{np}$ for all $\omega \in \Omega$;

(2.6) The process $\{X_{m+1,n+1}\}$ has the same joint distributions as the process $\{X_{mn}\}$;

(2.7) $E(X_{on}) < \infty$ for all $n \in N$ and $\inf_n E(\frac{X_{on}}{n}) \geq A$ for some constant A.

If $g_n \equiv E(X_{on})$, it follows from (2.6) that $E(X_{mn}) = g_{n-m}$, so that taking expectations in (2.5) gives $g_{p-m} \leq g_{n-m} + g_{p-n}$, implying that

(2.8) $g_{m+n} \leq g_n + g_m$ for all $m,n \in N$.

A standard result in the theory of subadditive functions ([29], p. 244) then implies that

(2.9) $\lim_{n\to\infty} g_n/n = \gamma$,

where

(2.10) $\gamma = \inf_{n \geq 1} g_n/n$

is finite because of (2.7). The constant γ will be called the "time constant" of the process.

The most powerful result about subadditive processes was proved by Kingman (1968):

> <u>Theorem 2.7</u> $\lim_{n\to\infty} \frac{X_{on}}{n}$ exists a.s. and in L^1.
> If ξ denotes this limit,
>
> $$E(\xi) = \gamma.$$
>
> If $\{X_{mn}\}$ is <u>independent</u> in the sense that whenever $n_1 < n_2 < \ldots < n_k$ the random variables $\{X_{n_i, n_{i+1}}\}_{i=1}^{k-1}$ are independent, then $\xi = \gamma$ a.s.

Several proofs are now known of this result; we give one in Appendix B.

In Chapter III a two-dimensional generalization of the idea of a subadditive process is used (cf. [18], [47]). In the two-dimensional case, we consider a process $\{X_{\underset{\sim}{mn}}: \underset{\sim}{m}, \underset{\sim}{n} \in N^2;\ \underset{\sim}{m} \leq \underset{\sim}{n}\}$ where $\underset{\sim}{m} = (m_1, m_2)$, $\underset{\sim}{n} = (n_1, n_2)$, and $\underset{\sim}{m} \leq \underset{\sim}{n}$ means that $m_1 \leq n_1$ and $m_2 \leq n_2$. Such a process will be called <u>subadditive</u> if it satisfies the three conditions below, analogous to (2.5)-(2.7):

(2.11) If $\underset{\sim}{m} < \underset{\sim}{p}$ and $m_2 < n_2 < p_2$, $X_{\underset{\sim}{mp}} \leq X_{\underset{\sim}{m}(p_1, n_2)} + X_{(m_1, n_2)\underset{\sim}{p}}$;

if $\underset{\sim}{m} < \underset{\sim}{p}$ and $m_1 < n_1 < p_1$, $X_{\underset{\sim}{mp}} \leq X_{\underset{\sim}{m}(n_1, p_2)} + X_{(n_1, m_2)\underset{\sim}{p}}$.

(2.12) The processes $\{X_{(m_1+1, m_2), (n_1+1, n_2)}\}$ and $\{X_{(m_1, m_2+1), (n_1, n_2+1)}\}$ have the same joint distributions as $\{X_{\underset{\sim}{mn}}\}$;

(2.13) $E(X_{o\underset{\sim}{n}}) < \infty$ for all $\underset{\sim}{n} \in N^2$ and $\inf_{\underset{\sim}{n}} E\left(\frac{X_{o\underset{\sim}{n}}}{|\underset{\sim}{n}|}\right) \geq A$

for some constant A.

The strict analogue of Theorem 2.7 does not hold for two-dimensional subadditive processes (cf. [46]), but convergence in mean has been established in [47]:

<u>Theorem 2.8</u> Let $\{X_{\underset{\sim}{mn}}\}$ be a subadditive process. Then $\gamma = \lim_{n} E\left(\frac{X_{o\underset{\sim}{n}}}{|\underset{\sim}{n}|}\right)$ exists, and $\gamma = \inf_{\underset{\sim}{n}} E\left(\frac{X_{o\underset{\sim}{n}}}{|\underset{\sim}{n}|}\right)$.

$\frac{X_{o\underset{\sim}{n}}}{|\underset{\sim}{n}|}$ converges in L^1 to a limit ξ, where $E(\xi) = \gamma$. If $\{X_{\underset{\sim}{mn}}\}$ is an independent subadditive process, $\xi = \gamma$ a.s.

2.4 The Kesten-Hammersley Theorem

In the discussion of [35] Kesten stated a useful result closely related to subadditive processes. This was soon extended by Hammersley (1974). We present a special case of Hammersley's result which suffices for our purposes.

Theorem 2.9 Let $\{X_n, n \in N\}$ be a sequence of random variables with distribution functions F_n and finite second moments. Suppose that for each pair (m,n) of positive integers with $m < n$ there exists a random variable X'_{mn} satisfying:

(i) X'_{mn} has distribution function F_n;

(ii) X_m and X'_{mn} are independent

(iii) $F_{n+m} \geq F_m * F_n$ for all m and n (here $*$ denotes the usual convolution operator)

Then there exists a constant γ such that

$$\frac{X_n}{n} \to \gamma \text{ in probability}$$

and for any $m \in N$,

$$\lim_{n\to\infty} \frac{X_{2^n m}}{2^n m} = \gamma \quad \text{a.s.}$$

If the sequence $\{X_n\}$ is monotone then

$$\frac{X_n}{n} \to \gamma \quad \text{a.s.}$$

Proof. Set

(2.14) $B_n = E(X_n)$, $C_n = \sqrt{E(X_n^2)}$, $D_n = C_n^2 - B_n^2 = \mathrm{Var}(X_n)$.

By conditions (i) - (iii) of the theorem,

(2.15) $B_{n+m} = E(X_{n+m}) \leq E(X_m + X'_{mn}) \leq B_m + B_n$

so that

(2.16) $\lim_n \frac{B_n}{n} = \gamma$ with $-\infty \leq \gamma < \infty$

Using the Minkowski inequality with conditions (i) and (iii), we have

(2.17) $C_{n+m}^2 = E(X_{n+m}^2) \leq E(X_m + X'_{mn})^2 \leq \{[E(X_m^2)]^{\frac{1}{2}} + [E(X'^2_{mn})]^{\frac{1}{2}}\}^2$

$\leq (C_m + C_n)^2$

Hence $0 \leq C_{n+m} \leq C_n + C_m$ and

(2.18) $\lim_n \frac{C_n}{n} = \theta$ where $0 \leq \theta < \infty$.

Now $B_n^2/n^2 \leq C_n^2/n^2$ so that from (2.16) and (2.18), $\gamma^2 \leq \theta^2 < \infty$.
Hence $\gamma > -\infty$.

By (2.14) and conditions (ii) and (iii),

(2.19) $D_{m+n} + B_{n+m}^2 = C_{n+m}^2 \leq E(X_m + X'_{mn})^2 = D_n + D_m + (B_m + B_n)^2$.

Let $m = n = 2^{r-1}t \equiv s(r-1)$ in (2.19) and set

$$(2.20) \qquad Y_r = \frac{B_{s(r)}}{s(r)} ;$$

then from (2.19),

$$(2.21) \qquad \frac{D_{s(r)}}{[s(r)]^2} \leq \frac{1}{2} \frac{D_{s(r-1)}}{[s(r-1)]^2} + Y_{r-1}^2 - Y_r^2$$

Note also that

$$(2.22) \qquad \frac{D_{s(1)}}{[s(1)]^2} = \frac{D_{2t}}{(2t)^2} \leq \frac{C_{2t}}{2t}^2 \leq \frac{C_{2t}}{2t}^2 + \frac{1}{2} \frac{D_{s(N)}}{[s(N)]^2} .$$

Summing (2.21) from $r = 2$ to $r = N$ and adding (2.22) to the result, we get

$$(2.23) \qquad \sum_{r=1}^{N} \frac{D_{s(r)}}{[s(r)]^2} \leq \frac{1}{2} \sum_{r=1}^{N} \frac{D_{s(r)}}{[s(r)]^2} + \left(\frac{C_{2t}}{2t}\right)^2 + Y_o^2 - Y_N^2$$

and hence

$$(2.24) \qquad \sum_{r=1}^{N} \frac{D_{s(r)}}{[s(r)]^2} \leq 2 \left(\frac{C_{2t}}{2t}\right)^2 + \left(\frac{B_t}{t}\right)^2 .$$

Since the right-hand side of (2.24) is independent of N, it follows that $\sum_{r=1}^{\infty} \frac{D_{s(r)}}{[s(r)]^2} < \infty$; by Chebyshev's inequality we have that for any $\varepsilon > 0$,

$$(2.25) \qquad \sum_{r=1}^{\infty} P\{|\frac{X_{s(r)}}{s(r)} - \frac{B_{s(r)}}{s(r)}| > \varepsilon \} < \infty$$

and the Borel-Cantelli lemma then gives

$$\frac{X_{s(r)}}{s(r)} - \frac{B_{s(r)}}{s(r)} \to 0 \text{ a.s. as } r \to \infty. \tag{2.26}$$

In view of (2.16), this proves that $\frac{X_{2^r t}}{2^r t} \to \gamma$ a.s. as $r \to \infty$, for any fixed t. Also, from (2.14), (2.16), and (2.18), and the convergence of $\sum_{r=1}^{\infty} \frac{D_{s(n)}}{[s(n)]^2}$,

$$\theta^2 - \gamma^2 = \lim_{n\to\infty} \frac{D_n}{n^2} = 0 \tag{2.27}$$

so that $\frac{X_n}{n} \to \gamma$ in probability and in L^2 as $n \to \infty$.

If $\{X_n\}$ is monotone the passage to a.s. convergence follows easily. Given $\epsilon > 0$, fix m so large that

$$\frac{m}{m+1} > 1 - \epsilon, \quad \frac{m+1}{m} < 1 + \epsilon. \tag{2.28}$$

For any t, there are positive integers n and k, with $1 \le k \le 2m$, such that $2^n(m+k-1) \le t < 2^n(m+k)$. Then by monotonicity,

$$\frac{(m+k-1)}{m+k} \frac{X_{2^n(m+k-1)}}{2^n(m+k-1)} \le \frac{X_t}{t} \le \frac{X_{2^n(m+k)}}{2^n(m+k)} \frac{m+k}{m+k-1} \tag{2.29}$$

and it follows by what has already been proved that

$$(1-\epsilon)\gamma \le \liminf_t \frac{X_t}{t} \le \limsup_t \frac{X_t}{t} \le (1+\epsilon)\gamma. \tag{2.30}$$

Since ϵ is arbitrary, it follows that $\frac{X_t}{t} \to \gamma$ a.s.

2.5 The connectivity constant

Let f_k denote the number of distinct self-avoiding paths of k bonds from the origin. The asymptotic behavior of f_k has been investigated by a number of authors for many other lattices as well as for the square lattice (see, for example, Hammersley (1961A, 1961B, 1963), Hammersley and Welsh (1962), Kesten (1963, 1964); the review paper of Shante and Kirkpatrick (1971) is a good source of references for this and related problems). The basic result which we will use was established by Hammersley (1957):

Theorem 2.10 $\lambda \equiv \lim_{n\to\infty} (f_n)^{\frac{1}{n}}$ exists.

Proof. Let $g_n = \frac{1}{n} \log f_n$; it suffices to show that g_n converges to a finite limit. It is easily seen that $f_{n+m} \leq f_n f_m$, and that $f_n \leq 3^n$; hence $\log f_n$ is a subadditive function, and it follows that $\lim_n g_n = \inf_n g_n = \mu$, say, where $0 \leq \mu < \infty$. Then $\lambda = e^{\mu}$.

The constant λ is called the connectivity constant of the square lattice. Fisher and Sykes (1959) have given bounds for λ and have estimated that $\lambda = 2.6395 \pm 10$. It is evident that a connectivity constant exists for other regular lattices as well; it is an important measure of the complexity of the path structures in the lattice (cf. [45]).

2.6 Two results of Hammersley

Hammersley (1966) proved two results which will be useful to us in establishing integrability properties of various processes and in the study of route length in Chapter VIII.

Let F be a distribution function on the half-line $[0,\infty)$.

Theorem 2.11 For $y \geq 0$, the function

$$\gamma(y) = \inf_{u \geq 0} \int_0^\infty e^{-u(x-y)} dF(x)$$

is non-decreasing and right continuous and satisfies

$$\gamma(0) = F(0) \leq F(y) \leq \gamma(y) \leq 1.$$

Proof. For $u \geq 0$, $y \geq 0$, let

$$\Phi(u,y) = \int_0^\infty e^{-u(x-y)} dF(x). \tag{2.31}$$

For $h \geq 0$,

$$\gamma(y) = \inf_{u \geq 0} \Phi(u,y) = \inf_{u \geq 0} e^{-hu} \Phi(u,y+h) \leq \inf_{u \geq 0} \Phi(u,y+h) \tag{2.32}$$

$$= \gamma(y+h)$$

Given $\epsilon > 0$, choose a finite non-negative $v = v(y,\epsilon)$ such that

$$\Phi(v,y) \leq \gamma(y) + \epsilon. \tag{2.33}$$

Then

$$(2.34)\qquad \gamma(y+h) \leq \Phi(v,y+h) = e^{hv}\Phi(v,y) \leq e^{hv}(\gamma(y)+\epsilon)$$

and it follows that

$$(2.35)\qquad \limsup_{h \to 0} \gamma(y+h) \leq \gamma(y) + \epsilon .$$

Since $\epsilon > 0$ is arbitrary, (2.32) and (2.35) show that $\gamma(y)$ is right continuous and non-decreasing for $y \geq 0$. Furthermore,

$$(2.36)\qquad \Phi(u,y) \geq \int_0^y e^{-u(x-y)}dF(x) \geq \int_0^y dF(x) = F(y).$$

Hence

$$(2.37)\qquad F(y) \leq \gamma(y) \leq \Phi(0,y) = 1.$$

Finally,

$$(2.38)\qquad \gamma(0) \leq \lim_{u\to\infty} \int_0^\infty e^{-ux}dF(x) = F(0);$$

from (2.37) and (2.38) we get

$$(2.39)\qquad \gamma(0) = F(0) \leq F(y) \leq \gamma(y) \leq 1.$$

Now consider a first-passage percolation process on the square lattice with non-negative time coordinates having distribution U. Let f_k denote the number of distinct self-avoiding paths of k bonds from the origin, and let m_k denote the minimum of the travel times over all these paths. Let $\gamma = \gamma_U$ as determined in Theorem 2.11.

<u>Lemma 2.12</u> For all $B > 0$ and all $k \in N$,

$$P(m_k \leq B) \leq f_k[\gamma(\tfrac{B}{k})]^k.$$

<u>Proof</u>. Let $U_1, U_2, \ldots$ be independent with the distribution U. Then $P(m_k \leq B) \leq f_k\ P(\sum_{i=1}^{k} U_i - B \leq 0) = f_k q_k$, say.

For any $u \geq 0$,

$$(2.40) \qquad q_k = P(\sum_{i=1}^{k} (U_i - {}^B/k) \leq 0) = P(-u \sum_{i=1}^{k} (U_i - \tfrac{B}{k}) \geq 0).$$

By Markov's inequality and independence,

$$(2.41) \quad P(-u \sum_{i=1}^{k} (U_i - \tfrac{B}{k}) \geq 0) \leq E(e^{-u \sum_{i=1}^{k}(U_i - \frac{B}{k})})$$

$$= [E\ (e^{-u(U_i - \frac{B}{k})})]^k$$

thus

$$q_k \leq \inf_{u>0} \left(\int_0^\infty e^{-u(x - \frac{B}{k})} dU(x)\right)^k = [\gamma(\tfrac{B}{k})]^k.$$

Chapter III - Bernoulli Percolation

Recall from Chapter I that the term "Bernoulli percolation" refers to the bond percolation model where the time coordinates are random variables assuming the values 0 and 1 only, with probabilities p and $q = 1 - p$, respectively. Bonds with travel time zero are called open; those with travel time one are closed. We envision a fluid flowing only along open bonds, wetting sites and bonds it reaches; among the problems we consider are those concerned with the size of connected clusters of open or wetted bonds in the medium.

As indicated in Chapter I, most work on percolation to date has concentrated on the Bernoulli model (site or bond). Our interest in this case is not just in Bernoulli percolation per se but also in the considerable illumination given by the Bernoulli model to problems in first-passage percolation. Given a general time coordinate distribution U we can, in a first-passage percolation model, regard a bond as open if its time coordinate is zero or less than some fixed number, and closed otherwise; results from Bernoulli percolation theory then allow us to draw some conclusions about first-passage percolation. Theorem 4.10 is a good example of this kind of application.

Much of this chapter follows the ideas of Seymour and Welsh, (1977), who proved the remarkable Theorem 3.8.

3.1 Definition of critical probabilities

Regard the fluid as being introduced at a particular source site. For small values of p, few bonds will be open, so the fluid will spread only locally before being completely blocked. For values of p near one, the fluid will penetrate throughout the medium. A fundamental question then arises: How large need p be for the fluid to penetrate the medium?

It seems intuitively reasonable to postulate that there is a threshold value above which penetration of the lattice occurs, and below which there is local spread only. This threshold concept is the basis for several definitions of a critical probability, depending on how one interprets the term "penetration of the lattice."

Let $\theta_n(p)$ be defined as the probability that at least n bonds of the lattice are wetted by fluid from the source site. Since obviously

$$\theta_n(p) \geq \theta_{n+1}(p) \quad \text{for all} \quad n,$$

the limit $\theta(p) = \lim_{n\to\infty} \theta_n(p)$ exists. $\theta(p)$ is called the <u>percolation probability</u> and represents the probability that the fluid spreads from the origin to an infinite set of bonds. Broadbent and Hammersley (1957) defined a <u>critical probability</u> $p_H = \inf\{p:\theta(p) > 0\}$. Although p_H is the most common critical probability appearing in the literature, several other definitions appear as well, and the relationships between them are far from clear.

One alternative, p_T, is the threshold value above which the expected number of bonds wet by the source site becomes infinite (see e.g. [57]). If $V(p)$ represents the expected size of this open cluster,

$$V(p) = \sum_{n=1}^{\infty} \theta_n(p)$$

and thus $p_T = \inf \{p : V(p) = \infty\}$.

Since $\theta(p) > 0$ implies $V(p)$ is infinite, it is clear that

$$p_T \leq p_H.$$

A third critical probability was defined for the sponge percolation model (3.21) by Seymour and Welsh as a technical convenience. Defining $S_n(p) = S_p(n, n+1)$, the probability of an open path crossing the $n \times (n+1)$ sponge, the critical sponge probability is defined as

$$p_S = \inf \{p : \limsup_{n\to\infty} S_n(p) > 0\}.$$

Harris (1960) proved the lower bound $p_H \geq \frac{1}{2}$. In the proof, a percolation probability $P(p, \frac{\pi}{2})$ was defined as the probability that there exists an infinite open cluster including the origin and contained in the first quadrant. We define a critical probability p_{H^+} by

$$p_{H^+} = \inf \{p : P(p, \tfrac{\pi}{2}) > 0\}.$$

Clearly $p_{H^+} \geq p_H$.

The relations between these various "critical probabilities" which will be presented in sections 3.2 - 3.7 may be summarized as follows:

$$\lambda^{-1} \leq p_T = p_S \leq \frac{1}{2} \leq p_H = p_{H^+} = 1 - p_T \leq 1 - \lambda^{-1},$$

where λ is the connectivity constant defined in §2.5.

Finally, Sykes and Essam (1964) defined a critical probability p_E as the location of a singularity of the function giving the mean number of clusters per bond on the lattice. Making at least two critical assumptions, Sykes and Essam showed that $p_E = \frac{1}{2}$ for the square lattice. This result has led to a "folklore" belief that the critical probability has been determined for the square lattice. Grimmett (1976) showed that one of the assumptions of Sykes and Essam is verifiable; we discuss this in §3.9. However, the nature of the definition of p_E is such that there is no known theoretical relationship between it and any of the other critical probabilities.

3.2 $p_T \leq p_S$

Seymour and Welsh established the following relation between the critical probabilities p_T and p_S:

Theorem 3.1 $p_T \leq p_S \leq \frac{1}{2}$.

Proof. By Theorem 2.2, $S_{\frac{1}{2}}(m,n) + S_{\frac{1}{2}}(n-1,m+1) = 1$; hence $S_{\frac{1}{2}}(n,n+1) + S_{\frac{1}{2}}(n,n+1) = 1$, and it follows that $S_n(\frac{1}{2}) = \frac{1}{2}$ for all n. Thus $p_S \leq \frac{1}{2}$.

To prove that $p_T \leq p_S$, we show that a finite expected cluster size implies that the sponge crossing probabilities $S_n(p)$ converge to zero.

Let T_n denote the $n \times (n+1)$ sponge, and let

$$R = \{(x,y):\ x = n+1,\ 1 \leq y \leq n\}$$

denote the set of vertices on the right border of T_n.

For each vertex $(1,i)$, $i \leq n$, let $|C_i(\omega)|$ denote the number of sites in the open bond cluster containing $(1,i)$. If $p < p_T$, the common expected cluster size $E|C_1|$ is finite. Therefore,

$$P[(1,i) \overset{T_n}{\rightsquigarrow} R] \leq P[|C_1| \geq n+1] \leq \frac{E[|C_1|\ ;\ |C_1| \geq n+1]}{n+1},$$

So

$$(3.1) \qquad S_n(p) = P[\overset{T_n}{\rightsquigarrow}] \leq P[\bigcup_{i=1}^{n} \{(1,i) \rightsquigarrow R\}] \leq \sum_{i=1}^{n} P[(1,i) \rightsquigarrow R]$$

$$\leq E[|C_1|\ ;\ |C_1| \geq n+1]$$

which converges to zero as $n \to \infty$. It follows that $p < p_S$; therefore $p < p_T$ implies $p < p_S$, so that $p_T \leq p_S \leq \frac{1}{2}$.

3.3 $p_T + p_H \leq 1$

Seymour and Welsh proved that $p_T + p_H \leq 1$; we give a different proof which establishes a stronger result.

Theorem 3.2 $p_T + p_H^+ \leq 1$.

Proof. We first obtain an inequality like (3.1) for the crossing probability $S_p(2n,n)$ of the sponge $T(2n,n)$.

If R again denotes the right border of $T(2n,n)$, then

$$S_p(2n,n) = P(\overset{T(2n,n)}{\rightsquigarrow})$$

$$= P(\bigcup_{i=1}^{2n} \{(1,i) \rightsquigarrow R\})$$

$$\leq \sum_{i=1}^{2n} P((1,i) \rightsquigarrow R) \leq \sum_{i=1}^{2n} P(|C_i| \geq n) = 2nP(|C_1| \geq n).$$

Consider now the following sequence of sponges in the first quadrant. If k is an even integer, B_k will denote the sponge $T(2^{k+1},2^k)$, whereas if k is odd, B_k will denote $T(2^k,2^{k+1})$. Thus each sponge is twice as long as it is wide, but the orientation with respect to the coordinate axes alternates with k.

Let E_k denote the event that B_k is crossed by a path of open bonds across its narrower dimension. We will show that $p < p_T$ implies

(3.2) $$P[E_k \text{ finitely often}] = 1.$$

To see this, consider

(3.3) $$\sum_{k=1}^{\infty} P(E_k) = \sum_{k=1}^{\infty} S_p(2^{k+1},2^k) \leq 2\sum_{k=1}^{\infty} 2^k P(|C_1| \geq 2^k).$$

For $2^{k-1} < i \leq 2^k$, we have $P(|C_1| \geq 2^k) \leq P(|C_1| \geq i)$. Since there are 2^{k-1} such integers i, we can bound the sum in (3.3) above by

$$4 \sum_{i=1}^{\infty} P(|C_1| \geq i);$$

this series converges if $p < p_T$, since $E|C_1|$ is then finite. By the Borel-Cantelli lemma, (3.2) follows.

By Corollary 2.3, for each B_k, either a path of open bonds crosses B_k in the narrow direction or a path of closed bonds crosses the dual of B_k lengthwise. According to (3.2), with probability one for all k sufficiently large the dual of B_k is crossed lengthwise by a path of closed bonds. By the construction of the boxes B_k, the closed lengthwise path in B_k^* crosses the width of B_{k+1}^*, for all k; it follows that with probability one, for all k sufficiently large the paths in B_k^* and B_{k+1}^* intersect. The union of these paths is therefore an infinite cluster of closed bonds contained in the first quadrant of the dual lattice. Now there exists k_o such that with probability at least one-half, $k > k_o$ implies that B_k^* is crossed lengthwise by a closed path. Since the probability that all bonds in the boxes $B_1^*, \ldots, B_{k_o}^*$ are closed is positive, it follows that the probability that $(\frac{1}{2}, \frac{1}{2})$ belongs to an infinite closed cluster in the dual is positive.

Therefore $q \geq p_{H^+}$, i.e.,

$$1 \geq p + p_{H^+},$$

and since this is true for all $p < p_T$,

$$1 \geq p_T + p_{H^+}\,.$$

3.4 The existence of circuits

By means of several lemmas in this section we will prove a lower bound on the probability of the existence of an open circuit in an annulus, which is a function of the sponge crossing probability. This result will be used in §3.5 to prove Harris' lower bound, and in §3.6 to establish the theorem of Seymour and Welsh.

Lemma 3.3 If $S(2n,2n) = \tau$, then $S(2n,4n) > \tau[1-(1-\tau)^{\frac{1}{2}}]^8$

Proof. Consider the following regions of the square lattice, which are illustrated in Figure 1:

$$R = \{(x,y):\ 1 \leq x \leq 4n,\ 1 \leq y \leq 2n\}$$
$$X = \{(x,y):\ x = 1,\ 1 \leq y \leq 2n\}$$
$$Z = \{(x,y):\ x = 2n,\ 1 \leq y \leq 2n\}$$
$$W = \{(x,y):\ x = n+1,\ 1 \leq y \leq 2n\}$$
$$W_1 = \{(x,y):\ x = n+1,\ 1 \leq y \leq n\}$$
$$U_1 = \{(x,y):\ n+1 \leq x \leq 3n,\ y = 1\}$$
$$U_2 = \{(x,y):\ n+1 \leq x \leq 3n,\ y = 2n\}$$
$$S_1 = \{(x,y):\ 1 \leq x \leq 2n,\ 1 \leq y \leq 2n\}$$
$$S = \{(x,y):\ n+1 \leq x \leq 3n,\ 1 \leq y \leq 2n\}$$

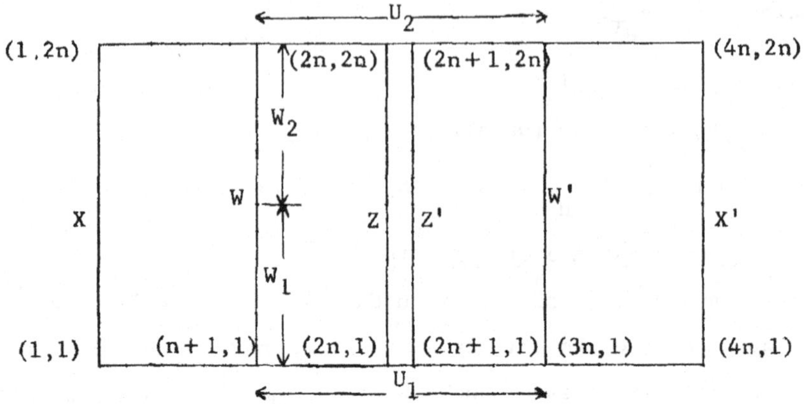

Figure 1

For any subset A of sites, let A' be defined as the reflection: $A' = \{(4n+1-x,y): (x,y) \in A\}$.

These three events will be considered:

$A_1 = \{\omega: W \overset{S}{\rightsquigarrow} W'\}$

$A_2 = \{\omega$: There is an open path from X to Z in S_1 which meets an open path from U_1 to U_2 in S .$\}$

$A_3 = \{\omega$: There is an open path from X' to Z' in S_1' which meets an open path from U_1 to U_2 in S .$\}$

The events A_1, A_2, and A_3 are monotone in the same sense, so by Theorem 2.4 they are covariant. Because

$$A_1 \cap A_2 \cap A_3 \subseteq \{X \overset{R}{\rightsquigarrow} X'\},$$

we have

$$(3.4) \qquad S(2n, 4n) \geq P(A_1 \cap A_2 \cap A_3) \geq P(A_1)[P(A_2)]^2$$
$$= S(2n,2n) \; [P(A_2)]^2$$

The remainder of the proof establishes the relation

$$(3.5) \qquad P(A_2) \geq [1 - (1 - \tau)^{\frac{1}{2}}]^4.$$

Let $\{P_i: 1 \leq i \leq k\}$ be the collection of paths in S_1 which join X to Z and which satisfy the additional property that their last point Q_i of intersection with W is a point of W_1. For $1 \leq i \leq k$, let F_i be the portion of P_i from Q_i to Z, so each F_i is a path from W_1 to Z.

Let X_i be the event that there is an open path in S from F_i to U_2 which uses only one site of F_i and no site of F_i'. Define X_i' as the event that an open path exists in S from F_i' to U_2 which uses only one site of F_i' and no site of F_i.

Since the set $F_i \cup F_i'$ separates U_1 from U_2 in S, if an open path from U_1 to U_2 exists in S, then either X_i or X_i' occurs. Hence

$$P[X_i \cup X_i'] \geq P[U_1 \overset{S}{\rightsquigarrow} U_2] = S(2n,2n) = \tau.$$

However, the events X_i and X_i' are covariant, and by symmetry their probabilities are equal. By Lemma 2.6,

$$P(X_i) = P(X_i') \geq 1 - \sqrt{1-\tau}$$

Fix i and consider the three events

$B_1 = B_1^{(i)} = \{\omega\colon$ the path P_i is open$\}$

$B_2 = B_2^{(i)} = \omega\colon$ for each $j \neq i$ such that P_j lies in the region bounded by P_i and $y = 1$, P_j is not open

$B_3 = B_3^{(i)} = X_i$.

We now claim that

$$(3.6) \qquad P(B_1 \cap B_2 \cap B_3) \geq [1 - \sqrt{1-\tau}]\, P(B_1 \cap B_2).$$

To see this, first condition on the event B_1:

$$(3.7) \qquad P(B_1 \cap B_2 \cap B_3) = P(B_2 \cap B_3 | B_1)\, P(B_1).$$

Notice that the occurrence of B_2 depends only on the state of edges strictly below P_i in S_1, whereas the occurrence of B_3 depends only on edges strictly above $F_i \cup F_i'$ in S. Thus, by conditional independence,

$$(3.8) \qquad P[B_2 \cap B_3 | B_1] = P[B_2|B_1]\, P[B_3|B_1].$$

Combining (3.7) and (3.8) we get

$$(3.9) \qquad P(B_1 \cap B_2 \cap B_3) = P(B_2|B_1)\, P(B_1)\, P(B_3|B_1) = P(B_1 \cap B_2)\, P(B_3|B_1$$

However, B_1 and B_3 are monotone in the same sense, so again by Theorem 2.4,

$$(3.10)\qquad P(B_3|B_1) \geq P(B_3) = P(X_i) \geq 1 - \sqrt{1-\tau}\ .$$

Substitution of (3.10) into (3.6) proves the claim.

Next, note that by disjointness,

$$P\left(\bigcup_{i=1}^{k} (B_1^{(i)} \cap B_2^{(i)})\right) = \sum_{i=1}^{k} P(B_1^{(i)} \cap B_2^{(i)}).$$

However,

$$(3.11)\qquad P\left[\bigcup_{i=1}^{k} (B_1^{(i)} \cap B_2^{(i)})\right] = P\left[\bigcup_{i=1}^{k} B_1^{(i)}\right],$$

and

$$(3.12)\qquad P\left[\bigcup_{i=1}^{k} (B_1^{(i)} \cap B_3^{(i)})\right] = \sum_{i=1}^{k} P[B_1^{(i)} \cap B_2^{(i)} \cap B_3^{(i)}]$$

$$\geq [1 - \sqrt{1-\tau}] \sum_{i=1}^{k} P(B_1^{(i)} \cap B_2^{(i)}) = [1 - \sqrt{1-\tau}]P(C),$$

where C is the event that at least one of the paths P_i is open.

Let $\{\hat{P}_i : 1 \leq i \leq k\}$ be the collection of paths in S_1 which join X to Z and which have the property that the last point of intersection with W is in W_2. Define $\hat{C}$ as the event that at least one of the paths $\hat{P}_i$ is open. The events C and $\hat{C}$ are covariant with equal probabilities, and

$$P(\hat{C} \cup C) = S(2n, 2n) = \tau.$$

Application of Lemma 2.6 yields

$$P(C) \geq 1 - \sqrt{1-\tau} ,$$

and therefore from (3.12),

$$(3.13) \qquad P(B_1^{(i)} \cap B_3^{(i)} \text{ occurs for some } i) \geq [1 - \sqrt{1-\tau}]^2 .$$

For the final step, define E_1 as the event that there exists a site $w \in W_1$ with $w \overset{S_1}{\rightsquigarrow} X$, $w \overset{S}{\rightsquigarrow} U_2$ and $w \overset{S}{\rightsquigarrow} Z$. Let E_2 be the event that there is a $v \in W_2$ such that $v \overset{S_1}{\rightsquigarrow} X$, $v \overset{S}{\rightsquigarrow} U_1$, and $v \overset{S}{\rightsquigarrow} Z$. The event E_1 occurs if, for some i, P_i is open and F_i is joined to U_2 by an open path in S, so

$$P(E_1) \geq P[\bigcup_{i=1}^{k} B_1^{(i)} \cap B_3^{(i)}] \geq [1 - \sqrt{1-\tau}]^2 ,$$

where the last inequality follows by (3.13).

But E_1 and E_2 are covariant, and $P(E_1) = P(E_2)$ by symmetry, so

$$P(E_1 \cap E_2) \geq P(E_2) \geq [1 - \sqrt{1-\tau}]^4 .$$

The event A_2 contains $E_1 \cap E_2$, so an application of (3.4) completes the proof.

<u>Lemma 3.4</u> $S(2n,6n) \geq [S(2n,2n)]^3 \, [1 - \sqrt{1 - S(2n,2n)}]^{16}$.

<u>Proof.</u> Consider the $2n \times 6n$ sponge and define the subregions

$$S = \{(x,y): 2n+1 \leq x \leq 4n,\ 1 \leq y \leq 2n\}$$

$$T = \{(x,y): 1 \leq x \leq 4n,\ 1 \leq y \leq 2n\}$$

$$U = \{(x,y): 2n+1 \leq x \leq 6n,\ 1 \leq y \leq 2n\} .$$

Let A be the event $\overset{T}{\rightsquigarrow}$, B the event $\overset{U}{\rightsquigarrow}$, C the event $\wr S$. Since $A \cap B \cap C \subseteq \overset{T(2n,6n)}{\rightsquigarrow}$, and A, B, and C are covariant events by Theorem 2.4,

$$S(2n,6n) \geq P(A)P(B)P(C) = [S(2n,4n)]^2 S(2n,2n),$$

and the conclusion now follows from Lemma 3.3.

Now let $R(n)$ denote the annular region of the square lattice bounded by the square C_n on the outside and D_n on the inside, where C_n consists of portions of the lines

$$y = -3n+1, \quad x = 3n, \quad y = 3n, \quad \text{and} \quad x = -3n+1$$

and D_n consists of portions of the lines

$$y = -n-1, \quad x = n+1, \quad y = n+1, \quad \text{and} \quad x = -n.$$

<u>Lemma 3.5</u> The probability that there exists an open circuit in the annulus $R(n)$ which encloses D_n and is enclosed by C_n is at least

$$S(2n,2n)^{12}[1 - \sqrt{1 - S(2n,2n)}]^{64}.$$

<u>Proof.</u> Define the following four subregions of the annulus:

$$\alpha = \{(x,y)\colon -3n+1 \leq x \leq -n, \ -3n+1 \leq y \leq 3n\}$$

$$\beta = \{(x,y)\colon -3n+1 \leq x \leq 3n, \ -3n+1 \leq y \leq -n-1\}$$

$$\Gamma = \{(x,y)\colon n+1 \leq x \leq 3n, \ -3n+1 \leq y \leq 3n\}$$

$$\Delta = \{(x,y)\colon -3n+1 \leq x \leq 3n, \ n+1 \leq y \leq 3n\}.$$

Let F_n be the event that there is an open circuit in $R(n)$ enclosing D_n. Then

$$F_n \supseteq \{\downarrow\alpha\} \cap \{\overset{\beta}{\rightsquigarrow}\} \cap \{\downarrow^{\Gamma}\} \cap \{\overset{\Delta}{\rightsquigarrow}\}.$$

The above four events are monotone in the same sense and thus covariant, so

$$P(F_n) \geq [S(2n,6n)]^4,$$

which with Lemma 3.4 completes the proof.

3.5 Harris' Lower Bound

The best theoretical lower bound for the critical probability p_H on the square lattice is $p_H \geq \frac{1}{2}$, established by Harris (1960). Harris' original proof is not presented here, but instead an argument based on Lemma 3.5. Both proofs depend on the symmetry when $p = \frac{1}{2}$ and on the almost sure existence of infinitely many circuits in disjoint annuli around the origin; they differ in the proof of existence of circuits.

Lemma 3.6 Suppose $p \geq \frac{1}{2}$. With probability one, infinitely many of the annuli $R(3^k)$ contain open circuits around the origin.

Proof. Suppose $p = \frac{1}{2}$. By Theorem 2.2 and the monotonicity property (2.1) of the sponge crossing probabilities,

$$S_{\frac{1}{2}}(2n,2n) \geq S_{\frac{1}{2}}(2n,2n+1) = \frac{1}{2} \quad \text{for all } n.$$

Let E_k be the event that an open circuit around the origin exists in the annulus $R(3^k)$. Then by Lemma 3.5,

$$P(E_k) \geq (\tfrac{1}{2})^{12}[1 - \sqrt{1-\tfrac{1}{2}}]^{64} > 0 \quad \text{for every } k.$$

Because the E_k are independent events with $\sum_{k=1}^{\infty} P(E_k)$ divergent, we conclude by the Borel-Cantelli lemma that

$$P(E_k \text{ infinitely often}) = 1.$$

Since this is true for $p = \frac{1}{2}$, it clearly holds for $p > \frac{1}{2}$ as well.

<u>Corollary 3.7</u> $\theta(\frac{1}{2}) = 0$; therefore $p_H \geq \frac{1}{2}$.

<u>Proof.</u> With probability one, $p = \frac{1}{2}$ implies that the origin of the dual lattice - the point $(\frac{1}{2},\frac{1}{2})$ - is enclosed by an open circuit in the lattice, hence (Theorem 2.1) by a cut set of open bonds in the dual lattice. Thus the probability is zero that $(\frac{1}{2},\frac{1}{2})$ is in an infinite cluster of closed bonds in the dual. Since the probability that each bond is closed is one-half, this shows that $\theta(\frac{1}{2}) = 0$, and it follows that $p_H \geq \frac{1}{2}$.

3.6 The Seymour-Welsh Theorem

An exact relation between the critical probabilities p_T and p_H was established by Seymour and Welsh (1977).

Theorem 3.8

$$p_T + p_H = 1, \; p_T = p_S \; \text{and} \; p_H = p_{H^+}.$$

We will prove Theorem 3.8 by a series of lemmas, in conjunction with the results of sections 3.2 and 3.3.

Lemma 3.9 If $\lim_n \sup S_p(n,n) > 0$, then $\lim_n \sup S_p(2n,2n) > 0$.

Proof. Suppose $\lim_n \sup S_p(n,n) > \varepsilon > 0$. Then by definition $S_p(n,n) > \epsilon$ for infinitely many n. The conclusion is obvious unless $S_p(n,n) > \epsilon$ holds for only finitely many even integers n. The remainder of the proof consists of demonstrating that, should this occur, then $S_p(2n,2n) > p\epsilon$ holds for infinitely many n.

Fix an odd integer n for which $S_p(n,n) > \epsilon$. By the monotonicity property (2.2) of sponge crossing probabilities,

$$S_p(n+1,n) \geq S_p(n,n) > \epsilon. \tag{3.14}$$

For each $i = 1,2,\ldots,n+1$, define E_i to be the event that an open path exists which crosses the sponge $T(n+1,n)$ to the site (n,i), and no open path crosses to (n,j) for any $j < i$.

For each $i = 1,2,\dots,n+1$, let B_i be the event that the bond from (n,i) to $(n+1,i)$ is open.

Clearly, if $E_i \cap B_i$ occurs, there is an open path crossing the $(n+1) \times (n+1)$ sponge, so

$$S_p(n+1,n+1) \geq P\Big(\bigcup_{i=1}^{n+1} (E_i \cap B_i)\Big).$$

Since the events E_i and B_i are independent, and $\bigcup_{i=1}^{n+1} E_i = \{T\xrightarrow{(n+1,n)}\}$, we conclude that

$$S_p(n+1,n+1) \geq \sum_{i=1}^{n+1} P(E_i \cap B_i) = p \sum_{i=1}^{n+1} P(E_i)$$

$$= p\ S_p(n+1,n) > p\ \epsilon$$

by (3.14). Now $n+1$ is an even integer, so

$$\limsup_{n\to\infty} S_p(2n,2n) \geq p \limsup_{n} S_p(n,n),$$

proving the lemma.

<u>Lemma 3.10</u> If $p < 1 - p_H$, then $\lim_{n\to\infty} S_p(n,n) = 0$.

Consequently, $p_S \geq 1 - p_H$.

<u>Proof.</u> Suppose $p < 1 - p_H$. Then $q > p_H$, implying that with positive probability there exists an infinite closed cluster containing the origin $(\frac{1}{2},\frac{1}{2})$ of the dual lattice.

Assume that $\limsup_{n\to\infty} S_p(n,n) > 0$. Then for some $\delta > 0$, by Lemma 3.9 we know that $S_p(2n,2n) > \delta$ for infinitely many n. Thus a sequence of integers $\{n_i\}$ exists for which $R(2n_i)$ are disjoint annuli and $S(2n_i, 2n_i) > \delta$ for each i.

By Lemma 3.5, there exists an open circuit in $R(2n_i)$ with probability greater than

$$\delta^{12}\ [1 - \sqrt{1-\delta}]^{64}$$

for each i.

These events are independent since the annuli $R(2n_i)$ are disjoint, so the Borel-Cantelli lemma implies that with probability one there can be no infinite closed cluster containing the origin in the dual lattice. This contradiction establishes that $\limsup_{n\to\infty} S_p(n,n) = 0$, so the limit exists and equals zero.

The next lemma is a strengthened version of a result of Seymour and Welsh, and it has consequences for Chapter VIII as well as for Theorem 3.8. We first define a "higher-moment" analogue of p_T.

For each positive integer m, let $V_m(p)$ denote the mth moment of the number of bonds in the open cluster containing the origin, when p is the probability of each bond being open.

For each m, define a critical probability $p_T m$ by

$$p_T m \equiv \inf \{p: V_m(p) = \infty\}.$$

(Evidently $p_T 1$ coincides with the p_T defined previously.)

<u>Lemma 3.11</u> If $\varepsilon > 0$ and $p > p_T m$, then for infinitely many values of k,

$$[1 - S_p(2k,2k)]^{12}\ [1 - \sqrt{S_p(2k,2k)}]^{64} \le 1 - \frac{1}{3^{2m}} + \varepsilon. \tag{3.15}$$

Proof. Fix m, and suppose the lemma were false. We could then choose N such that for all $k \geq 3^N$, the inequality (3.15) is reversed. By (2.1), (2.2), and Theorem 2.2 (setting $q = 1 - p$):

(3.16) $$S_q(2k,2k) \geq S_q(2k-1,2k+1) = 1 - S_p(2k,2k).$$

Applying Lemma 3.5 we deduce that, for $k \geq 3^N$, the probability that there is a closed circuit in the annulus $R(k)$ is at least $1 - \frac{1}{3^{2m}} + \epsilon$. By Theorem 2.1, such a closed circuit corresponds to a cut set of closed bonds in the dual lattice; so the event D_k that there is an open path from the origin through $R(k)$ in the dual has probability at most $\frac{1}{3^{2m}} - \epsilon$.

Let $|C|$ denote the number of bonds in the open cluster containing the origin in the dual lattice. By looking at the sequence of disjoint annuli $R(3^k)$, we obtain

$$E(|C|^m) \leq (4 \times 3^{2N})^m + \sum_{n \geq N} (4 \times 3^{2n})^m \prod_{k=N}^{n-1} P(D_k)$$

$$\leq 4^m \times 3^{2Nm} + \sum_{n=N}^{\infty} 4^m \times 3^{2nm} \left(\frac{1}{3^{2m}} - \epsilon\right)^{n-N} < \infty,$$

which contradicts $p > p_T m$.

A statement equivalent to Lemma 3.11 is that $p > p_T m$ implies $\limsup_{n \to \infty} S_p(2n,2n) \geq \delta > 0$. Seymour and Welsh note that in the case $m = 1$, $\delta \geq 5 \times 10^{-6}$.

Corollary 3.12 $p_{T^m} \geq p_S$ for all m.

Proof. By Lemma 3.11, $p > p_{T^m}$ implies $\limsup_{n\to\infty} S_p(2n,2n) \geq \delta > 0$. Reasoning as in the proof of Lemma 3.9, we may conclude that

$$\limsup_{n\to\infty} S_n(p) \geq \limsup_{k\to\infty} S_p(2k,2k+1) \geq p\delta > 0.$$

Hence $p_{T^m} \geq p_S$ for all m.

It is not known whether $S(p) = \lim_{n\to\infty} S_n(p)$ exists for all p. Seymour and Welsh note, however, that since

$$p > p_T \text{ implies } \limsup_n S_n(p) \geq \delta$$

and $$p < p_T \text{ implies } \lim_n S_n(p) = 0,$$

the limit $S(p)$, if it exists, is certainly discontinuous at p_T.

To complete the proof of Theorem 3.8, notice that Theorem 3.1 and Corollary 3.12 together show that $p_S = p_{T^m}$ for all $m \geq 1$. From Theorem 3.2 and Lemma 3.10,

$$p_T + p_{H^+} \leq 1 \leq p_T + p_H$$

Since $p_{H^+} \geq p_H$, we conclude that $p_{H^+} = p_H$ and $p_T + p_H = 1$.

As a consequence of the proof of Theorem 3.8, we have that $p_T = p_{T^m}$ for all m. Thus (except possibly when $p = p_T$), either all moments of order greater than or equal to one of the cluster size are finite, or they are all infinite. This observation verifies what appears to be a folk-lore belief that there is a critical probability at which all moments simultaneously become infinite, and identifies this value as p_T.

3.7 Hammersley's upper bound for p_H

Hammersley (1959) proved the upper bound for p_H given below. It is now an immediate consequence of Theorems 2.10 and 3.8.

> Theorem 3.13 $p_H \leq 1 - \lambda^{-1}$,
> where λ is the connectivity constant defined in §2.5. Hence (taking λ less than or equal to the upper bound 2.6405),
> $$p_H \leq .621.$$

Proof. In view of Theorem 3.8 it suffices to prove that $\lambda^{-1} \leq p_T$. Let ℓ be the length of the longest self-avoiding path of open bonds from the origin, and choose $p < \lambda^{-1}$. Then

$$P(\ell \geq n) \leq \sum_{k=n}^{\infty} f_k p^k,$$

where f_k, defined in §2.5, is the number of distinct k-bond self-avoiding paths from (0,0) . By Theorem 2.10, there exists $A > 0$ such that

$$f_k \leq A\lambda^k \quad \text{for all } k,$$

so that

$$\sum_{k=n}^{\infty} f_k p^k \leq A(\lambda p)^n (1-\lambda p)^{-1}.$$

Hence

$$\sum_{n=1}^{\infty} n\, P(\ell \geq n) < \infty,$$

So $E(\ell^2) < \infty$. If $|C|$ is the number of bonds in the open cluster containing the origin, it is easy to see that

$$|C| \geq n \quad \text{implies} \quad \ell^2 \geq n.$$

Hence $E|C| < \infty$, so that $p \leq p_T$; thus $\lambda^{-1} \leq p_T$.

3.8 The number of infinite clusters

Harris (1960) made the following interesting observation concerning the number of infinite zero clusters:

> Theorem 3.14 If $p < p_H$ the probability of there being an infinite open cluster anywhere in L is zero.
>
> If $p > p_H$ then, with probability one, there is exactly one infinite open cluster, and no infinite closed cluster.

Proof. The first statement has already been made in the proof of Theorem 3.2. If $p > p_H$, the construction of Theorem 3.2 produces, with probability one, an infinite open cluster; so it remains only to be shown that there is no other infinite cluster.

Let $\underset{\sim}{m}$ and $\underset{\sim}{n}$ be sites in L, and V any finite rectangular region of L such that $\underset{\sim}{m}$ and $\underset{\sim}{n}$ are both in the interior of V.

From the proof of Lemma 3.6, $p > p_H$ implies that, with

probability one, there is an open circuit enclosing V. It follows that, again with probability one, an infinite zero cluster containing $\underset{\sim}{m}$ is joined by zero bonds to any infinite open cluster containing $\underset{\sim}{n}$; the theorem is proved.

Fisher (1961) has shown that in any two-dimensional site percolation problem, there cannot be both an infinite cluster of open sites and an infinite cluster of closed sites.

3.9 The number of clusters per bond and mean cluster size

We next discuss the result of Grimmett alluded to in §3.1 and some related questions. Given a finite region of the lattice L, the number of open clusters per bond is simply the total number of distinct open clusters in the region divided by the number of bonds. Both of these quantities in general tend to infinity as the region expands; one of the assumptions made by Sykes and Essam (1964) was that the expected number of open clusters per bond converges to a limiting value as the region expands. (The limiting value will of course depend on p, the size of the atom at zero.) Grimmett's result provides a rigorous justification of this assumption.

Actually, Sykes and Essam (and Grimmett as well) considered the number of clusters per <u>site</u>; in keeping with our interest in first-passage percolation, we consider bonds instead. The analysis is not essentially different in the two cases, but in comparing our results with, for example, those of Welsh (1977), it should be borne in mind that in our definition of (bond) cluster a cluster containing a given site or bond may have size zero, whereas his definition of (site) cluster requires a cluster to have at least one site. Our function $\theta(p)$ below is thus not strictly comparable to his $\lambda(p)$.

Let R_{mn} be the rectangular region in the first quadrant bounded by the lines $x = 0$, $x = m$, $y = 0$, $y = n$. The bonds on the lines $x = m$ and $y = n$ will not be included in R_{mn}, but those on $x = 0$ and $y = 0$ will. There are thus $2mn$ bonds in R_{mn}.

<u>Definition 3.15</u> $C_{mn}(i) \equiv$ the number of open clusters of size i contained in R_{mn}.

$C(i) \equiv$ the number of open clusters of size i in the entire lattice which have at least one bond in R_{mn}.

If B is a particular bond,

$N_{mn}(B) \equiv$ the number of bonds in R_{mn} which are in the open cluster containing B ($N_{mn}(B) = 0$ if B is closed.)

$N(B) \equiv$ the number of bonds in the open cluster containing B.

$$C_{mn} \equiv \sum_{i=1}^{2mn} C_{mn}(i), \quad C \equiv \sum_{i=1}^{\infty} C(i).$$

Note that an open cluster in the lattice L may give rise to two or more distinct open clusters in R_{mn}, but each open cluster in R_{mn} corresponds to just one open cluster in L. Hence

(3.17) $\quad C(\omega) \leq C_{mn}(\omega)$ for all ω.

A weak form of Grimmett's result, in the bond formulation, is given in Theorem 3.16 below:

Theorem 3.16 $\quad \lim_{m,n\to\infty} \frac{C_{mn}}{2mn} = \theta(p)$ in L^r

for any $r > 0$, where

$$\theta(p) = \inf_{m,n} \frac{E(C_{mn})}{2mn}.$$

If k_{mn} is the total number of open clusters in the region bounded by the lines $x = \pm m$, $y = \pm n$ (including the bonds on $x = -m$ and $y = -n$ but not those on $x = m$ and $y = n$), then

$\frac{k_{mn}}{8mn} \to \theta(p)$ in L^r as $m,n\to\infty$, for any $r > 0$.

Proof. If $\underset{\sim}{m} = (m_1,m_2) \leq (n_1,n_2) = n$, let $C_{\underset{\sim}{mn}}$ be the number of zero clusters in the rectangle bounded by the lines $x = m_1$, $x = n_1$, $y = m_2$, $y = n_2$ (with the boundary convention of Definition 3.15 .

The argument leading to (3.17) shows that $\{C_{mn}\}$ satisfies the subadditivity condition (2.11). The stationarity condition (2.12) is obvious, and since $0 \leq C_{mn} \leq 2mn$, condition (2.13) is equally obvious. Because $\{C_{mn}\}$ is clearly an independent process, Theorem 2.8 applies to show that there is a constant $\theta = \theta(p)$, with $0 \leq \theta(p) \leq 1$, such that

$$(3.18) \qquad \frac{C_{mn}}{2mn} \to \theta(p) \quad \text{in } L^1 \quad \text{as } m,n\to\infty,$$

and $\theta(p) = \inf_{m,n} \frac{E(C_{mn})}{2mn}$. The boundedness of $\frac{C_{mn}}{2mn}$ then shows that convergence takes place in any mean. The extension to k_{mn} is accomplished by dividing the rectangle bounded by $x = \pm m$ and $y = \pm n$ into four isomorphic parts, by cutting along the coordinate axes. Each subadditive process $C^1_{mn}, C^2_{mn}, C^3_{mn}, C^4_{mn}$ in the four parts then has the same distribution, and, if N is the number of open bonds with exactly one site on the coordinate axes inside the large rectangle,

$$(3.19) \qquad \sum_{i=1}^{4} C^i_{mn} - N \leq k_{mn} \leq \sum_{i=1}^{4} C^i_{mn}.$$

The first inequality in (3.19) holds because any bond which contributes to N belongs to exactly one of the k_{mn} clusters in the large rectangle, and the removal of this bond increases the number of such clusters by at most one; the second inequality is obvious. Since $N \leq 4(m+n)$, $\frac{k_{mn}}{8mn}$ converges in any mean to $\theta(p)$, by (3.18).

Remark. Grimmett in fact proved a.s. convergence of $k_{mn}/8mn$ to the constant $\theta(p)$; since we have no need of the stronger result we omit the details.

We proceed now to study the function $\theta(p)$ of Theorem 3.16. Grimmett proved that $p \to \theta(p)$ is continuous, and that θ has a derivative equal to one at $p = 0$. We give below a characterization of $\theta(p)$ which allows us to prove much more:

Theorem 3.17 $\theta(p) = \sum_{i=1}^{\infty} \frac{1}{i} P[N(B) = i]$, where $N(B)$ is given in Definition 3.15.

The function $p \to \theta(p)$ is continuous and differentiable almost everywhere (Lebesgue) in p. We have $\theta'(0) = 1$, $\theta'(1) = 0$, and $\theta'(p) \leq 1$ for all p for which the derivative exists.

Finally, $\theta(p) \leq -(1-p) \ln(1-p)$ for all $p \in [0,1]$.

Proof. We note first that C_{mn} cannot exceed C by more than the number of open bonds in the boundary of R_{mn}, where the boundary of R_{mn} is defined to consist of all bonds lying outside of the rectangle defined by the lines $x = 0$, $x = m$, $y = 0$, and $y = n$ which have exactly one site on the boundary of this rectangle, plus the bonds along the boundaries $x = m$ and $y = n$

of this rectangle. (Recall that these last-mentioned bonds were not included in R_{mn}.) Thus

(3.20) $$C_{mn}(\omega) \leq C(\omega) + (3m+3n+4).$$

By (3.17) we have

(3.21) $$C_{mn} \geq C = \sum_{i=1}^{\infty} C(i) \geq \sum_{i=1}^{\infty} \frac{1}{i} \sum_{B\in R_{mn}} 1[N(B) = i]$$
$$= \sum_{B\in R_{mn}} \sum_{i=1}^{\infty} \frac{1}{i} 1[N(B) = i]$$

so that

(3.22) $$\frac{E(C_{mn})}{2mn} \geq \sum_{i=1}^{\infty} \frac{1}{i} P(N(B) = i).$$

Next note that

(3.23) $$C_{mn} = \sum_{i=1}^{2mn} C_{mn}(i) = \sum_{i=1}^{\infty} \frac{1}{i} \sum_{B\in R_{mn}} 1[N_{mn}(B) = i]$$
$$= \sum_{B\in R_{mn}} \sum_{i=1}^{\infty} \frac{1}{i} 1[N_{mn}(B) = i].$$

Fix ω. For any bond B in a cluster which is totally contained in R_{mn},

(3.24) $$\sum_{i=1}^{\infty} \frac{1}{i} 1[N_{mn}(B) = i] = \sum_{i=1}^{\infty} \frac{1}{i} 1[N(B) = i].$$

The bonds contained in clusters which are not entirely in R_{mn} can form no more clusters than there are open bonds in the boundary of R_{mn}, since each cluster must leave R_{mn} through some open bond in the boundary. For bonds of this type, if ∂R denotes the boundary of R_{mn},

$$(3.25) \qquad \sum_{B} \sum_{i=1}^{\infty} \frac{1}{i} 1_{[N_{mn}(B) = i]} \leq \sum_{B \in \partial R} 2\, 1_{[B \text{ is open}]},$$

since the sum of the left-hand side over bonds in any particular R_{mn}-cluster is one, but one or more of the open bonds in ∂R serves as an exit from R_{mn}; and a single open boundary bond can not be an exit for more than two R_{mn}-clusters. From (3.23) and (3.25), therefore,

$$(3.26) \qquad R_{mn} \leq \sum_{B \in R} \sum_{i=1}^{\infty} \frac{1}{i} 1_{[N(B) = i]} + 2 \sum_{B \in \partial R} 1_{[B \text{ is open}]}, \text{ and}$$

using (3.20),

$$(3.27) \qquad \frac{E(C_{mn})}{2mn} \leq \sum_{i=1}^{\infty} \frac{1}{i} P[N(B) = i] + \frac{2p(3m + 3n + 4)}{2mn}$$

Letting $m, n \to \infty$ in (3.22) and (3.27), we get by Theorem 3.16 that

$$(3.28) \qquad \sum_{i=1}^{\infty} \frac{1}{i} P[N(B) = i] = \theta(p).$$

Then from (3.27):

$$\frac{E(C_{mn})}{2mn} - \theta(p) \leq \frac{p(3m + 3n + 4)}{mn},$$

so that convergence of $\frac{E(C_{mn})}{2mn}$ to $\theta(p)$ is uniform in p; since $E(C_{mn})$ is a polynomial in p for all p, it follows that $p \to \theta(p)$ is a continuous function.

Next, observe that by Abel summation in (3.28),

$$(3.29) \qquad \theta(p) = p - \sum_{i=2}^{\infty} \frac{1}{i(i-1)} P_p(N(B) \geq i)$$

But $P_p(N(B) \geq i)$ is monotone increasing in p, so (3.29) shows that $\theta(p)$ is differentiable a.e. with $\theta'(p) \leq 1$. For p near zero, the series in (3.29) is clearly $o(p)$, so $\theta'(0) = 1$; since $P_p(N(B) \geq 2) = p - p(1-p)^6$, $\frac{\theta(p)}{1-p}$ is $o(1-p)$ as $p \to 1$, so $\lambda'(1) = 0$. Thus although $\theta(0) = \theta(1) = 0$, the function $p \to \theta(p)$ is not symmetric about $p = \frac{1}{2}$.

Finally we observe that $P_p[N(B) \geq i] \geq p^i$ for any p and i, so by (3.29),

$$(3.30) \qquad \theta(p) \leq p - \sum_{i=2}^{\infty} \frac{1}{i(i-1)} p^i$$

An easy calculation gives the bound $\theta(p) \leq -(1-p)\ln(1-p)$, so that in particular:

$$(3.31) \qquad \theta(p) \leq e^{-1} \quad \text{for all } p \in [0,1].$$

Sykes and Essam (1964) speak of a "singularity" in the average number of clusters per site (as a function of p) at the critical value p_E defined in §3.1; in light of the results above for the bond function $\theta(p)$ it is not clear what "singularity" should be taken to mean.

We turn finally to the expected cluster size at a particular bond B, i.e., $E(N(B))$. Of course, we must assume that $p < p_T$ to make sense of the following discussion. Seymour and Welsh (1977), in their discussion of cluster size, confirm one's intuition by proving that the "average cluster size" is not greater than the expected size of the cluster through a given site. A similar result holds for bond clusters, properly interpreted. If

$$O_{mn} \equiv \text{the number of open bonds in } R_{mn},$$

then

O_{mn}/C_{mn} can be thought of as the average cluster size in R_{mn}, and

$$\lim_{m,n\to\infty} \frac{O_{mn}}{C_{mn}} = \lim_{m,n\to\infty} \frac{O_{mn}}{2mn}\frac{2mn}{C_{mn}} = \frac{p}{\theta(p)} \quad \text{in probability,} \tag{3.32}$$

by the law of large numbers and Theorem 3.16 (note that, by (3.28), $\theta(p) \neq 0$ for $p \neq 0$ or 1). Hence $\frac{p}{\theta(p)}$ is the "average cluster size" for the bond problem.

Theorem 3.18 For a fixed bond B,
$E(N(B)) \geq \frac{p^2}{\theta(p)}$; in other words,
$E[N(B) | N(B) \geq 1] \geq \frac{p}{\theta(p)}$, the average

cluster size.

Proof. $1_{[N(B) > 0]} = \sqrt{N(B)} \left(\frac{1}{\sqrt{N(B)}} 1_{[N(B) > 0]}\right)$ and by the Cauchy-Schwarz inequality,

$$(3.17) \qquad p \leq \{E(N(B)\; E[\frac{1}{N(B)} 1_{[N(B) > 0]}]\}^{\frac{1}{2}}$$

Using Theorem 3.16, $\theta(p) = E[\frac{1}{N(B)}; N(B) > 0]$ so that (3.33) gives

$$E[N(B)] \geq \frac{p^2}{\theta(p)}.$$

In the next section we note that it is widely believed that $p_H = \frac{1}{2}$. If this is the case, it is clearly of interest to know the expected cluster size when $p = \frac{1}{2}$; it is in fact infinite, as we prove below.

Theorem 3.19 If $p = \frac{1}{2}$, the expected cluster

size $E[N(B)]$ is infinite.

Proof. As noted in the proof of Theorem 3.1, $S_n(\frac{1}{2}) = \frac{1}{2}$ for all n, where $S_n(\frac{1}{2})$ is the probability of crossing the $n \times (n+1)$ sponge from left to right on open bonds.

Fix n and for $i = 0,1,\dots,n-1$, let E_i be the event that an open path leads from the point $(1,i)$ to the right-hand edge of the sponge; let $|C_i|$ denote the size of the open cluster at $(1,i)$. Since $S_n(\frac{1}{2}) = \frac{1}{2}$, it follows that

$$(3.34)\qquad P(E_i) \geq \frac{1}{2n} \quad \text{for some } i \in \{0,1,\dots,n-1\}.$$

But $\quad E_i \subset \{|C_i| \geq n\}$ for each i; so

$$(3.35)\qquad P(|C_i| \geq n) \geq \frac{1}{2n} \quad \text{for some } i \in \{0,1,\dots,n-1\}.$$

However, $|C_i|$ has the same distribution as the size of the open cluster through some fixed point, say $(0,0)$; and if B denotes the bond from $(0,0)$ to $(1,0)$, $E(N(B))$ is at least one-fourth the size of this last cluster. Hence

$$P(N(B) \geq n) \geq \frac{1}{8n} \quad \text{for every } n$$

and it follows that $E[N(B)] = \infty$.

Note that, in contrast with the result of Theorem 3.19, the average cluster size $\frac{p}{\theta(p)}$ is finite for all values of p. Temperley and Lieb (1971) have shown that when $p = \frac{1}{2}$, the mean number of sites in a cluster is at least 10.204.

3.10 Monte Carlo Results

Shante and Kirkpatrick (1971) list the results of a number of computer simulation studies aimed at estimating various critical probabilities. Some care is necessary in determining exactly which critical probability is being considered in these studies; in particular, the results listed as "exact" determinations in [45] refer to the values p_E of Sykes and Essam, discussed in §3.1. The results of these studies lend strong support to the long-standing conjecture that $p_T = \frac{1}{2}$; by the results of this chapter, a proof of this result would have the pleasing consequence that all of the critical probabilities for the square lattice have the value one-half.

More recent numerical studies can be found in Kurkijarvi and Padmore (1975) (for the site process), and in a series of papers of Sykes, Gaunt, and Glen (1976). Kurkijarvi and Padmore based their studies on a site version of the sponge model discussed in Chapter II. Other useful results which are not mentioned by Shante and Kirkpatrick include those of Dean and Bird (1967).

Chapter IV

Definition of the basic processes and the existence of routes

This chapter will introduce the basic first-passage processes (§4.1) and study their integrability properties (§4.2). In §4.3 we will show that routes (as defined in §1.2) exist a.s. for these processes.

4.1 The basic first-passage processes

In this section we will define the processes that form our principal object of study. We first define the "unrestricted processes."

> <u>Definition 4.1</u> Suppose that $m < n$. Let R_{mn} be the set of all self-avoiding paths from the point $(m,0)$ to the point $(n,0)$.
>
> $a_{mn}(\omega) \equiv t_{R_{mn}}(\omega)$ is called the <u>first-passage time</u> from $(m,0)$ to $(n,0)$.
>
> Let $\tilde{R}_{mn}$ be the set of all self-avoiding paths from the point $(m,0)$ to some point on the line $x = n$, which are contained entirely, except for the last end point, in $\{x < n\}$.
>
> $b_{mn}(\omega) \equiv t_{\tilde{R}_{mn}}(\omega)$ is called the <u>first-passage time</u> from $(m,0)$ to the line $x = n$.

By Lemma 1.1 we have $b_{mn} \leq a_{mn}$ for all $m < n$, if the time coordinates are non-negative. In Definition 4.1, the restriction of the paths to be self-avoiding is essential if we wish to consider negative time coordinates; otherwise the possibility of negative loops would give $-\infty$ as the minimal travel time. Here, as later, the restriction of $\tilde{R}_{mn}$ to paths in $\{x < n\}$ is made to prevent ambiguities that may arise if the paths are allowed to travel the line $x = n$.

In their analysis of a_{mn} and b_{mn}, Hammersley and Welsh found it convenient (indeed, essential) to consider other point-to-point and point-to-line times where the set of possible paths is restricted. If $m < n$, a <u>cylinder path</u> is defined to be a path contained entirely, except for its first and last endpoints, in the cylinder $m < x < n$. (This is a slight variant of Hammersley and Welsh's definition in [27]; the processes t_{on} and s_{on} defined below do not correspond exactly to theirs, but are asymptotically equivalent.) The "cylinder analogues" of a and b are the processes below:

<u>Definition 4.2</u> Suppose $m < n$. Let C_{mn} be the set of all self-avoiding cylinder paths from the point $(m,0)$ to the point $(n,0)$.

$t_{mn}(\omega) \equiv t_{C_{mn}}(\omega)$ is called the <u>cylinder first-passage time</u> from $(m,0)$ to $(n,0)$.

Let $\tilde{C}_{mn}$ be the set of all self-avoiding cylinder paths from the point $(m,0)$ to some point (n,k) on the line $x=n$.

$s_{mn}(\omega) \equiv t_{\tilde{C}_{mn}}(\omega)$ is called the <u>cylinder first-passage time</u> from $(m,0)$ to the line $x=n$.

It follows from Lemma 1.1 that $a_{mn} \leq t_{mn}$, $b_{mn} \leq s_{mn}$, and that $s_{mn} \leq t_{mn}$.

In analyzing the first-passage processes above, several other processes will be useful. If $m<n$, consider the set of all paths from a fixed point on the line $x=m$ to some point on the line $x=n$ which satisfy the following: the first step is either up, down, or to the right; the rest of the path (except possibly the first, and certainly the last, endpoint) is contained in the cylinder $m<x<n$. Such paths will be called <u>three-way cylinder paths</u>. (A three-way cylinder path thus has three possible first steps, whereas a cylinder path has only one.)

<u>Definition 4.3</u> Let Z_{mn} be the set of all self-avoiding three-way cylinder paths from $(m,0)$ to some point (n,k) on the line $x=n$.

$$z_{mn}(\omega) \equiv t_{Z_{mn}}(\omega)$$

We then have $b_{mn} \leq z_{mn} \leq s_{mn}$ for all m and n.

Finally we consider a class of processes whose paths are restricted to a half-plane.

> <u>Definition 4.4</u> Let D^k_{mn} be the set of all self-avoiding paths from $(1,m)$ to $(1,n)$ which are entirely contained in $\{x \geq 1-k\}$, for $k > 0$.
>
> $$d^k_{mn}(\omega) \equiv t_{D^k_{mn}}(\omega)$$
>
> (when $k = 0$ we will suppress the superscript and write simply d_{mn}).

All of the above processes have been defined for passage in one preferred direction. In Chapter V we shall need to consider, for example, first-passage times from $(0,0)$ to $(0,n)$; since the time coordinates U_i are i.i.d., this passage time has the same distribution as a_{on} and we denote it $a_{on}(\uparrow)$. Likewise $a_{on}(\downarrow)$ is the first-passage time from $(0,0)$ to $(0,-n)$, when $n > 0$. A similar notation will be used for the processes d of Definition 4.4; thus $d_{on}(\uparrow)$ is just the d_{on} defined above, whereas $d_{on}(\downarrow)$ is the first-passage time from $(1,0)$ to $(1,-n)$, where $n > 0$, over paths in $\{x \geq 1\}$.

The processes $a,b,s,$ and t are our primary objects of study z and d will be used only as intermediaries in two proofs.

4.2 Integrability results for first-passage processes

In this section we show the relationship of moments for the time coordinate distribution U to moments of the first-passage processes; throughout the section it will be assumed that the time coordinates are non-negative with finite mean. It will be seen that the cylinder processes have much weaker integrability properties than the unrestricted processes. (The results of this section are taken from [60].)

For fixed $n > 0$, we consider travel times on each of four disjoint paths from $(0,0)$ to $(n,0)$. The first path is straight along the x-axis. A second path begins with a vertical step to $(0,1)$, then continues horizontally along the line $y = 1$ to $(n,1)$, then steps down to $(n,0)$. The third path is the reflection of the second path across the x-axis. The fourth path consists of travel to $(-1,0)$, $(-1,2)$, $(n+1,2)$, $(n+1,0)$, then $(n,0)$, along straight-line segments between successive sites in this sequence. The travel time of the ith path will be denoted by $t(i,n)$ for $1 \leq i \leq 4$.

We begin with the cylinder processes.

> Theorem 4.5 If the time coordinate distribution U has a finite m^{th} moment, for $m \geq 1$, then s_{on} and t_{on} have finite m^{th} moments for all n, and the processes
>
> $$\left\{\frac{s_{on}}{n}\right\}^m \quad \text{and} \quad \left\{\frac{t_{on}}{n}\right\}^m$$
>
> are uniformly integrable.

If the mth moment of U does not exist, then for all $n \geq 1$ the mth moments of s_{on} and t_{on} do not exist.

Proof. By definition

$$s_{on} \leq t_{on} \leq t(1,n).$$

But $t(1,n)$ is just the nth partial sum of an i.i.d. sequence with distribution U, so $\{\frac{t(1,n)}{n}\}^m$ is a reversed submartingale and given $\epsilon > 0$,

$$E\{\left(\frac{t_{on}}{n}\right)^m; \left(\frac{t_{on}}{n}\right)^m > \lambda\} \leq E\{U_1^m; \left(\frac{t_{on}}{n}\right)^m > \lambda\}$$

$$< \epsilon \quad \text{for all } n \text{ if } \lambda > \lambda_o.$$

Uniform integrability of $\left(\frac{t_{on}}{n}\right)^m$ and $\left(\frac{s_{on}}{n}\right)^m$ follows.

By the cylinder restriction, $t(1,1) \leq s_{on} \leq t_{on}$ for all $n \geq 1$, and since $t(1,1)$ is distributed as U, the second statement follows.

We turn next to the unrestricted first-passage times.

Theorem 4.6 Let U_1, U_2, U_3, U_4 be independent with the time coordinate distribution U. For any $m \geq 1$,

If $E(\min_{1 \leq i \leq 4} U_i^m) < \infty$, then $E(a_{on}^m)$ and $E(b_{on}^m)$ are finite for all $n \geq 1$;

If $E(\min_{1 \leq i \leq 4} U_i^m) = \infty$, then $E(a_{on}^m) = E(b_{on}^m) = \infty$ for all $n \geq 1$.

Proof. By definition, $b_{on} \leq a_{on} \leq \min_{1\leq i\leq 4} t(i,n)$ for all n.

Therefore

(4.1) $$E(a_{on}^m) < \infty \quad \text{if} \quad E(\min_{1\leq i\leq 4} (t(i,n)^m) < \infty,$$

which is equivalent to the condition

(4.2) $$\sum_{k=1}^{\infty} P(\min_{1\leq i\leq 4} (t(i,n)^m > k) < \infty.$$

Since the four paths being examined are disjoint, the times $t(i,n)$ are independent for $1 \leq i \leq 4$. Therefore

(4.3) $$P(\min_{1\leq i\leq 4} (t(i,n))^m > k) = \prod_{i=1}^{4} P((t(i,n))^m > k)$$

Consider the fourth path; for any n, by labelling the time coordinates on the fourth path as $\{U_i\}_{i=1}^{n+8}$, beginning at the origin, we have

(4.4) $$P((t(4,n))^m > k) = P((U_1 + \ldots + U_{n+8}) > k^{\frac{1}{m}})$$

$$\leq (n+8)P(U_1 > k^{\frac{1}{m}}/n+8).$$

A similar argument applies to each of the other three paths. Letting U_i, $1 \leq i \leq 4$, denote the time coordinates of the first bond on the ith path:

(4.5) $$P(\min_{1\leq i\leq 4} (t(i,n)^m > k) \leq (n+8)^4 P(\min_{1\leq i\leq 4} U_i^m > (n+8)^{-m}k).$$

Hence if $E(\min_{1\leq i\leq 4} U_i^m) \leq \infty$, it follows by (4.1), (4.2), and (4.5) that $E(a_{on}^m) < \infty$.

The second assertion of the theorem follows from the fact that any path for a_{on} or b_{on} must leave the origin on one of four bonds, which have independent time coordinates with distribution U. □

<u>Lemma 4.7</u> Let $\{X(i,n)\}$, $1 \leq i \leq 4$, be independent, uniformly integrable processes. Then the process $\{\min_{1\leq i\leq 4} X(i,n)^4\}$ is uniformly integrable.

<u>Proof</u>. $\sup_n E(\min_{1\leq i\leq 4} X(i,n)^4;\ \min_{1\leq i\leq 4} X(i,n)^4 > \lambda)$

$$\leq \sup_n E(\prod_{i=1}^{4} X(i,n);\ X(i,n) > \lambda^{\frac{1}{4}}$$

$$= \prod_{i=1}^{4} \sup_n E(X(i,n);\ X(i,n) > \lambda^{\frac{1}{4}})$$

which tends to zero as $\lambda\to\infty$ by the uniform integrability of the $X(i,n)$, $1\leq i\leq 4$.

<u>Theorem 4.8</u> If the time coordinate distribution U has a finite m^{th} moment, where $m \geq 1$, then the processes

$$\{\frac{b_{on}}{n}\}^{4m} \quad\text{and}\quad \{\frac{a_{on}}{n}\}^{4m}$$

are uniformly integrable.

Proof. We apply Lemma 4.7 to the processes $\{\frac{t(i,n)}{n}\}^m$, for $1 \leq i \leq 4$.

$\left(\frac{t(4,n)}{n}\right)^m$ is dominated by $\{9^m \left(\frac{S_{n+8}}{n+8}\right)^m\}$ for each n, where, in the notation of Theorem 4.6, $S_{n+8} = U_1 + \ldots + U_{n+8}$. But $\{\frac{S_n}{n}\}^m$ is a reversed submartingale and therefore uniformly integrable by the argument used in Theorem 4.5; it follows that $\left(\frac{t(4,n)}{n}\right)^m$ is uniformly integrable. A similar argument holds for $t(2,n)$ and $t(3,n)$; by (4.1) and Lemma 4.3, $\left(\frac{a_{on}}{n}\right)^m$ is uniformly integrable.

Remark It may happen that U has infinite m^{th} moment but $\min(U_1, U_2, U_3, U_4)$ has finite $4m^{th}$ moment (take $P(U > k) \sim c/k^m \log k$). However, if the $4m^{th}$ moment of $\min(U_1, U_2, U_3, U_4)$ is finite, U must have moments of all orders less than m.

> Theorem 4.9 If the time coordinate distribution U has a finite m^{th} moment, where $m \geq 1$, then for some constant $k(m,U)$,
> $$E\left(\frac{z_{on}}{n}\right)^{3m} \leq k(m,U) \quad \text{for all } n \geq 1.$$

Proof. We proceed as in Theorem 4.6, using only the first three paths described at the beginning of the section; the proof is then completed as in Theorem 4.8.

4.3 Existence of routes

Routes for first-passage times were defined in §1.2. Here we will prove that routes exist with probability one for all the first-passage times defined in §4.1. The fact that routes exist will allow us, in Chapter V, to select a path with travel time equal to the first-passage time, eliminating some bothersome epsilonics. Also, in Chapter VIII, we need to know that routes exist in order to consider properties of the routes, such as the number of bonds they contain and the maximum height they attain.

4.3.1 Existence of routes for non-negative time coordinates

All time coordinates in this sub-section will be assumed to be non-negative. Hammersley and Welsh (1965) verified the existence of routes when the time coordinates are bounded above and below away from zero. For unbounded distributions, they proved that routes exist a.s. for t_{on} and s_{on}. The almost sure existence of routes for b_{on} for all distributions was established in [49], and for a_{on} for all distributions with atom at zero not equal to the critical probability p_H. This final case was handled in [61].

A common feature in proofs of existence of routes is the concept of a barrier, which is a boundary such that any path crossing this boundary has a travel time not less than the first-passage time. The barrier restricts consideration to the finite collection of paths which do not cross it, and the path(s) among those with the minimum travel time is then a route.

<u>Theorem 4.10</u> Routes exist for a_{on}, b_{on}, d_{on}, s_{on}, t_{on}, and z_{on} for all n with probability one, for every time coordinate distribution U which does not charge $(-\infty,0)$.

<u>Proof</u>. Fix a positive integer n. We will consider separately the cases $U(0) < \frac{1}{2}$ and $U(0) \geq \frac{1}{2}$, finding a different type of barrier for each situation.

Suppose first that $U(0) \geq \frac{1}{2}$. Consider bonds with zero travel time to be open, and those with positive travel time to be closed. Let $\tilde{\Omega}$ be the event that an open circuit exists which contains both the origin and the point $(n,0)$ in its interior. Lemma 3.6 guarantees that an infinite number of such circuits exists with probability one, so $P(\tilde{\Omega}) = 1$.

For $\omega \in \tilde{\Omega}$, there is an open circuit C enclosing a region containing both $(0,0)$ and $(n,0)$. There are only a finite number of self-avoiding paths in the region enclosed by C, and among these paths there is, for each of the first-passage times a_{on}, b_{on}, d_{on}, s_{on}, t_{on}, and z_{on}, at least one path with a minimal travel time.

We claim that C acts as a barrier, i.e., that for any of the processes, there is a route in the region enclosed by C. For a point-to-point first-passage time, this is true because a path leaving the enclosed region must re-enter the region to reach its destination, and the distance from the exit point to the re-entry point may be traveled in zero time along the circuit. For a point-to-line first-passage time, a path which leaves the region and travels to the line $x = n$ will have travel time at least as great as the path which travels (in zero time) the open circuit from the exit point to the intersection of the open circuit with the line $x = n$. Therefore, consideration may be restricted to paths in the region, and any of these with minimal travel time is a route.

Now suppose that $U(0) < \frac{1}{2}$, so there exists $x > 0$ for which $U(x) < \frac{1}{2}$. Regard bonds with travel time less than x as open, and bonds with travel time greater than or equal to x as closed. Let L^* be the dual lattice (cf. §2.1), where a bond is assigned the same time coordinate as the bond it crosses, so the probability that a bond is closed in the dual lattice is at least one-half. By Lemma 3.6, we can, with probability one, find a square B_k in L^*, centered at $(\frac{1}{2}, \frac{1}{2})$, so large that there are k disjoint annuli inside B_k, each containing a closed circuit around $(\frac{1}{2}, \frac{1}{2})$; let Ω_k be the set of sample points for which this is true.

Let R_n be the event that routes of a_{on}, b_{on}, d_{on}, s_{on}, t_{on}, and z_{on} all exist, and let S_n denote the travel time from the origin to $(n,0)$ along the x-axis.

Suppose $\omega \in \Omega_k$. Then any path from the origin to the boundary of B_k crosses k closed circuits in the dual, and so contains at least k bonds of travel time x or larger. It follows that any path to the boundary of B_k has travel time at least kx, so

$$\{S_n \leq kx\} \cap \Omega_k \subseteq R_n \quad \text{for all } k.$$

Finally, since $P(\Omega_k) = 1$,

$$P(R_n) \geq \lim_{k\to\infty} P(S_n \leq kx) = 1.$$

Therefore $P(R_n) = 1$, so $P(\bigcap_n R_n) = 1$, i.e., routes exist a.s. for all n.

4.3.2 Existence of routes with negative time coordinates

In Chapter VIII we shall need to consider time coordinate distributions which may take small negative values. If r is a real number, let $\omega \oplus r$ denote the time state of the lattice obtained by adding the constant r to the time coordinate of each bond. Even if the original distribution U is positive, the shifted distribution function, viz.,

$$U \oplus r(x) = U(x-r) \quad \text{for } x \text{ real}$$

may take on positive values at negative x, if $r < 0$. This seems heretical if one thinks of the usual percolation models in which the time coordinate represents the time taken for a particle to traverse a bond; the extension is nonetheless technically useful and may even be justified physically in a routing problem where a "bonus" might accrue from passing through a particular town or station.

If routes are to exist in this case, it is clear that some restriction is needed on the negative part of the distribution U. For example, if U has a negative mean, we can construct a path for t_{o2} by traveling first to $(1,0)$, then vertically to $(1,k)$, horizontally to (2,k), then down to (2,0). By the law of large numbers, the infimum over k of the travel times for these paths is $-\infty$, but no path attains $-\infty$ as its travel time.

The previous route existence proof, for non-negative time coordinates, constructed barriers which could not be crossed without exceeding the travel time of some other path. When time coordinates are allowed to be negative, the travel time over a portion of a path may exceed the travel time over the entire path, so the barrier arguments no longer apply. (The λ below was defined in §2.5.)

> Theorem 4.11 Let U be a non-negative time coordinate distribution with $U(0) < \lambda^{-1}$, and suppose that $r < 0$ satisfies $\gamma(-r) < \lambda^{-1}$. Then routes exist almost surely, for all n, for $a_{on}(\omega\oplus r)$, $b_{on}(\omega\oplus r)$, $d_{on}(\omega\oplus r)$, $s_{on}(\omega\oplus r)$, $t_{on}(\omega\oplus r)$, $z_{on}(\omega\oplus r)$.

Proof. The same proof works for all the first-passage times, so we give it only for $a_{on}(\omega\oplus r)$.

Fix $B > 0$ and let

(4.6) $$R_B \equiv \sup\left\{\begin{array}{l}\text{number of bonds in a self-avoiding path from} \\ (0,0) \text{ with travel time less than } B\end{array}\right\}$$

Let m_i be defined as in §2.6, i.e., m_i is the infimum of the travel times over all self-avoiding paths of i bonds from the origin. Then

$$\{R_B(\omega\oplus r) > k\} \subseteq \bigcup_{i=k}^{\infty} \{m_i(\omega\oplus r) < B\}$$

By Lemma 2.12,

$$P\{m_i(\omega\oplus r) < B\} = P\{m_i(\omega) < B-ir\} \leq f_i[\gamma(\tfrac{B}{i}-r)]^i.$$

Choose $c<1$ such that $\gamma(-r)<c^2\lambda^{-1}$ and choose i_o so large that

$$\gamma(\tfrac{B}{i_o}-r) < c^2\lambda^{-1}$$

and $f_i < (\lambda/c)^i$ for all $i \geq i_o$.

Then for $k \geq i_o$,

(4.7) $$P\{R_B(\omega\oplus r) > k\} \leq \sum_{i=k}^{\infty} c^i = c^k(1-c)^{-1}.$$

Since there are a finite number of self-avoiding paths from $(0,0)$ with fewer than k bonds, for ω in the event

(4.8) $$\{R_B(\omega\oplus r) \leq k\} \cap \{a_{on}(\omega\oplus r) < B\},$$

a route exists for $a_{on}(\omega\oplus r)$. By choosing B and k sufficiently large, the probability of the complement of the event (4.8) can be made arbitrarily small, by (4.7). Thus a route exists for $a_{on}(\omega\oplus r)$ with probability one, so routes exist a.s. for all n.

Chapter V

Convergence of the first-passage processes

This chapter is devoted to establishing the fundamental convergence results which are used throughout the sequel. Sections 5.1 and 5.2 - 5.3 treat, respectively, the point-to-point and point-to-line processes; in §5.4 the convergence results are extended to allow negative values for the time coordinates. In an attempt to minimize confusion, the symbol U will hereafter be used only for distributions which do not charge $(-\infty,0)$ and which have finite mean.

§5.1 Convergence of $\frac{t_{on}}{n}$ and $\frac{a_{on}}{n}$.

Consider the point-to-point processes t_{mn} and a_{mn} defined in Chapter IV. Because a path from $(m,0)$ to $(p,0)$ may, but need not, pass through $(n,0)$ when $m<n<p$, Lemmas 1.1 and 1.2 show that t and a both satisfy condition (2.5) in the definition of a subadditive process. Since the time coordinates are i.i.d., (2.6) holds also; and we may take $A=0$ in (2.7) because the processes are non-negative. A "weak law" for these processes was given by Hammersley and Welsh (1965); invoking Theorem 2.7 we can prove the following "strong law" for t_{on} and a_{on}:

<u>Theorem 5.1</u> Let the time coordinates have distribution U. There exists a non-negative constant $\mu(U)$ such that

$$\lim_n \frac{t_{on}}{n} = \mu(U) \text{ a.s.}$$

$$\lim_n \frac{a_{on}}{n} = \mu(U) \text{ a.s.}$$

<u>Proof</u>. The result for t_{on} follows directly from Theorem 2.7, noting that the cylinder restriction makes t_{mn} an independent process. The process a_{mn} is not independent, however, so all that can be concluded from Theorem 2.7 is that $\frac{a_{on}}{n}$ converges a.s. to a limit ξ, where $\xi \leq \mu(U)$ a.s. since $a_{on} \leq t_{on}$.

Let $\mu_A(U)$ be the "time constant" for the process a_{mn}, i.e., $E(\xi) = \mu_A(U)$. Clearly $\mu_A(U) \leq \mu(U)$, so if we can prove $\mu_A(U) \geq \mu(U)$, Theorem 5.1 will be established.

We follow here the proof of Hammersley and Welsh ([27], p. 87-88) Take $m < n$ and k any positive integer. Let $q^k_{mn}(\omega)$ be the first-passage time between $(m,0)$ and $(n,0)$ under ω, over paths in the infinite strip $m-k < x \leq n+k$. By Theorem 4.10, routes exist a.s. for q^k_{mn}, for all k. By Lemmas 1.1 and 1.2, stationarity, and the non-negativity of q, $\{q^k_{mn}\}$ is a subadditive process for any fixed k.

Let $Q_k(n) \equiv E(q^k_{on})$ and let $\mu_k(U)$ be the time constant of the q^k-process. Then by (2.9) and (2.10),

$$(5.1)\qquad \frac{Q_k(n)}{n} \geq \mu_k(U) = \lim_n \frac{Q_k(n)}{n} .$$

It is evident from Lemma 1.1 that

$$(5.2)\qquad a_{on} \leq q_{on}^{k} \leq q_{on}^{k-1} \leq t_{on} \quad (k \geq 2),$$

so that

$$(5.3)\qquad \mu_A(U) \leq \mu_k(U) \leq \mu_{k-1}(U) \leq \mu(U) \quad (k \geq 2).$$

Let r_o be a route of $t_{-k,o}(\omega)$, r_1 a route of $q_{on}^{k}(\omega)$, r_2 a route of $t_{n,n+k}(\omega)$. Then $r_o * r_1 * r_2$ is a connected cylinder path from $(-k,0)$ to $(n+k,0)$ and hence

$$(5.4)\ t_{-k,n+k}(\omega) \leq t(r_o * r_1 * r_2,\omega) = t(r_o,\omega) + t(r_1,\omega) + t(r_2,\omega)$$
$$= t_{-k,o}(\omega) + q_{on}^{k}(\omega) + t_{n,n+k}(\omega).$$

Taking expectations in (5.4), and letting $\tau(n) = E(t_{on})$,

$$(5.5)\qquad \tau(n+2k) \leq 2\tau(k) + Q_k(n).$$

Dividing (5.5) by n and letting $n\to\infty$ with k fixed, we get

$$(5.6)\qquad \mu(U) \leq \mu_k(U)$$

which with (5.3) implies that

$$(5.7)\qquad \mu_k(U) = \mu(U) \quad \text{for all } k.$$

Now for n,ω fixed, $q_{on}^{k}(\omega)$ is decreasing in k, and

$$(5.8)\qquad \lim_k q_{on}^{k}(\omega) = a_{on}(\omega) .$$

By monotone convergence,

$$(5.9)\qquad \lim_k Q_k(n) = E(a_{on}) \quad (n \text{ fixed}) .$$

But by (5.1) and (5.7),

$$(5.10) \qquad \frac{Q_k(n)}{n} \geq \mu(U) \quad \text{for all } k,n.$$

By (5.9) and (5.10),

$$(5.11) \qquad E\left(\frac{a_{on}}{n}\right) \geq \mu(U)$$

and letting $n\to\infty$ in (5.11), $\mu_A(U) \geq \mu(U)$.

5.2 Convergence of $\frac{s_{on}}{n}$

The point-to-line processes s_{mn} and b_{mn} defined in section 4.1 are not subadditive, so that direct application of Theorem 2.7 is not possible. Let $\psi(n) = E(s_{on})$; we can at least show that $\psi(n)$ is a subadditive function.

Lemma 5.2 $\psi(m+n) \leq \psi(n) + \psi(m)$.

Proof. Let r_1 be a route of $s_{on}(\omega)$ and let it meet the line $x=n$ at $p \equiv (n,y_1)$. Let $s^{y_1}_{n,m}$ be the first-passage time from p to the line $x=m+n$ over cylinder paths whose first bond joins (n,y_1) to $(n+1,y_1)$. Clearly $s^{y_1}_{n,m}$ has the same distribution as s_{om}; and by Lemma 1.1,

$$(5.12) \qquad s_{on} + s^{y_1}_{n,m} \geq s_{o,m+n}.$$

Take expected values in (5.12); then

$$E(s^{y_1}_{n,m}) = \sum_{k=-\infty}^{\infty} E(s^k_{n,m} | y_1 = k)P(y_1 = k) = \psi(m) \sum_{k=-\infty}^{\infty} P(y_1 = k) = \psi(m),$$

and the result follows.

Let $\mu_s(U) \equiv \lim_n \frac{\psi(n)}{n}$; since $s_{on} \leq t_{on}$, it follows that

(5.13) $\mu_s(U) \leq \mu(U)$.

In fact equality holds in (5.13), as the next theorem shows:

<u>Theorem 5.3</u> $\mu_s(U) = \mu(U)$.

<u>Proof</u>. We follow the proof of Hammersley and Welsh ([27], pp. 83-85). Let k be a positive integer, and let $s_k^1(\omega) = s_{ok}(\omega)$. Select a route at random - call it r_1 - from the (possibly singleton) set of routes having the smallest maximum vertical displacement from the x-axis, and suppose r_1 meets the line $x = k$ at $P_1 \equiv (k, h_1(\omega))$. Let $s_k^2(\omega)$ be the cylinder first-passage time from $(k, h_1(\omega))$ to the line $x = 2k$ and let r_2 be a randomly chosen route of s_k^2 with the smallest maximum vertical deviation from the line $y = h_1(\omega)$; let r_2 meet the line $x = 2k$ at $P_2 \equiv (2k, h_1(\omega) + h_2(\omega))$. Continuing in this way we define sequences

$$\{s_k^i(\omega)\}_{i=1}^n, \ \{r_i\}_{i=1}^n, \ \{h_i(\omega)\}_{i=1}^n.$$

such that

a) $\{s_k^i(\omega)\}_{i=1}^n$ is an i.i.d. sequence with the distribution of $s_{ok}(\omega)$;

b) $\{r_i\}_{i=1}^n$ is a sequence of paths such that $r_1 * r_2 * \ldots * r_n$ is a connected path from the origin to the line $x = nk$;

c) $\{h_i(\omega)\}_{i=1}^n$ is a sequence of i.i.d., symmetric, integer-valued random variables.

Define $H_n(\omega) \equiv \sum_{i=1}^{n} h_i(\omega)$ and let r_o be the straight line path from $(nk, H_n(\omega))$ to $(nk, 0)$, followed by one step to the right. If h_i^2 is integrable, it will follow that H_n^2 is integrable and that

(5.14) $$(E|H_n|)^2 \leq E(H_n^2) = \sum_{i=1}^{n} E(h_i^2) = nE(h_i^2) .$$

Lemma 5.4 $E(h_i^2) < \infty$.

Proof. Let A_p denote the set of all bonds in the rectangle defined by $0 \leq x \leq k$, $p \leq y \leq p+k$, except for the horizontal bonds on the line $y = p+k$. We say that A_p forms a barrier if every bond in the horizontal segment $\{0 \leq x \leq k,\ y = p\}$ has a time coordinate smaller than the smallest time coordinate in the rest of A_p. The route r_1 cannot cross A_p vertically if A_p is a barrier, because the time needed to do this would be greater than the passage time along the segment $\{0 \leq x \leq k,\ y = p\}$. Unless the time coordinate distribution is degenerate (in which case Theorem 5.3 is obvious), the probability that A_p is a barrier for any $p \geq 0$ is strictly positive, say π_o. If $h_1(\omega) \geq kj$, then none of the sets $A_o, A_k, \ldots, A_{(j-1)k}$ can have been a barrier; and since these sets are disjoint, the probability of this does not exceed $(1-\pi_o)^j$. Hence

$$P(h_1(\omega) \geq kj) \leq (1-\pi_o)^j$$

and since h_1 is symmetric, it follows that $E(h_1^2) < \infty$.

Now by Lemma 1.1 and b) above, if w_n denotes the time coordinate of the bond joining $(nk,0)$ with $(nk+1,0)$,

$$(5.15)\qquad t_{o,nk+1}(\omega) \le t(r_1 * r_2 * \ldots * r_n * r_o, \omega)$$

$$= \sum_{i=1}^{n} t(r_i,\omega) + t(r_o,\omega) = \sum_{i=1}^{n} s_k^i(\omega) + t(r_o,\omega) + w_n.$$

By the independence of H_n and the time coordinates on the line $x = nk$, $E[t(r_o,\omega)] = \bar{u}\, E|H_n|$, where $\bar{u}$ is the mean of the time coordinate distribution. Hence, taking expectations in (5.15) and using (5.14):

$$(5.16)\qquad \tau(nk+1) \le n\,\psi(k) + \bar{u}\, n^{\frac{1}{2}}\{E(h_1^2)\}^{\frac{1}{2}} + \bar{u}.$$

Dividing by $nk+1$, when $n\to\infty$ we get

$$(5.17)\qquad \mu(U) \le \frac{\psi(k)}{k}.$$

Now let $k\to\infty$; by definition of $\mu_s(U)$ we have

$$(5.18)\qquad \mu(U) \le \mu_s(U).$$

With (5.13), this completes the proof of Theorem 5.3.

The corollary below was established by Hammersley and Welsh (1965):

<u>Corollary 5.5</u> $\frac{s_{on}}{n} \to \mu(U)$ in probability as $n\to\infty$.

<u>Proof</u>. By Theorem 5.3, $E(\frac{t_{on}-s_{on}}{n}) \to 0$ as $n\to\infty$, so $\frac{t_{on}-s_{on}}{n} \to 0$ in probability. Since $\frac{t_{on}}{n} \to \mu(U)$ a.s. by Theorem 5.1, it follows that $\frac{s_{on}}{n} \to \mu(U)$ in probability.

In [61] the stronger result that $\frac{s_{on}-t_{on}}{n} \to \mu(U)$ a.s. was proved. This almost follows from Theorem 2.9 except that s_{on} need not be square-integrable as required by that result (cf. Theorem 4.1). To get around that problem we work instead with z_{on}, defined in §4.1.

Select a route for the process z_{on}. We define $z'_{n,m}$ to be the first-passage time from the endpoint of the route of z_{on} to the line $x = m+n$ over three-way cylinder paths. Clearly $z'_{n,m}$ has the same distribution as z_{om}.

<u>Lemma 5.6</u> z_{on} and $z'_{n,m}$ are independent.

<u>Proof</u>. Let $P(\omega) = i$ if (n,i) is the endpoint of the route of $z_{on}(\omega)$. For fixed i, let $z^i_{n,m}$ be the first-passage time from (n,i) to the line $x = n+m$ over three-way cylinder paths. For any non-negative numbers α and β,

$$P(z_{on} \leq \alpha,\ z'_{n,m} \leq \beta) = \sum_{i=-\infty}^{\infty} P(z_{on} \leq \alpha,\ P = i,\ z^i_{n,m} \leq \beta).$$

By independence and stationarity,

$$P(z_{on} \leq \alpha,\ P = i,\ z^i_{n,m} \leq \beta) = P(z_{on} \leq \alpha, P = i)\ P(z^i_{n,m} \leq \beta)$$

$$= P(z_{on} \leq \alpha,\ P = i)\ P(z_m \leq \beta)$$

for each i, so that

$$P(z_{on} \leq \alpha,\ z'_{n,m} \leq \beta) = P(z_m \leq \beta) \sum_{i=-\infty}^{\infty} P(z_{on} \leq \alpha,\ P = i)$$

$$= P(z'_{n,m} \leq \beta)\ P(z_{on} \leq \alpha).$$

Theorem 5.7 $\lim_n \frac{z_{on}}{n} = \mu(U)$ a.s.

Proof. We invoke Theorem 2.9; evidently z_{on} is a monotone sequence and Theorem 4.9 shows that the z_{on} have finite second (in fact, third) moments. By definition of $z'_{n,m}$ and Lemma 5.6, conditions (i) and (ii) of Theorem 2.9 are satisfied; (iii) holds because

$$z_{o,m+n} \leq z_n + z'_{n,m}.$$

Hence $\frac{z_{on}}{n}$ converges a.s. to a constant $\mu_z(U)$, where $\mu_z(U) \leq \mu(U)$ because $z_{on} \leq t_{on}$.

For given i, let $s_{on}(i)$ denote the cylinder first-passage time from (0,i) to the line $x = n$. Each $s_{on}(i)$ thus has the same distribution as s_{on}. Let U^+ and U^- denote the time coordinates of the bonds from (0,0) to (0,1) and (0,-1) respectively, and note that

$$z_{on} = \min\{s_{on}, s_{on}(1) + U^+, s_{on}(-1) + U^-\}.$$

By Corollary 5.5, $\frac{s_{on}}{n} \to \mu(U)$ in probability; hence $\frac{z_{on}}{n} \to \mu(U)$ in probability, and therefore $\mu_z(U) = \mu(U)$.

We come at last to the main result of this section:

Corollary 5.8 $\frac{s_{on}}{n} \to \mu(U)$ a.s. as $n \to \infty$.

<u>Proof</u>. We have $z_{on} \leq s_{on} \leq t_{on}$; by Theorem 5.1, $\frac{t_{on}}{n} \to \mu(U)$ a.s., and by Theorem 5.7, $\frac{z_{on}}{n} \to \mu(U)$ a.s.

5.3 Convergence of $\frac{b_{on}}{n}$ and mean convergence

It was shown in [48] that $\frac{b_{on}}{n} \to \mu(U)$ in probability when the distribution U is bounded above and below away from zero; these bounds were removed in [49]. Finally in [61] it was shown that convergence holds a.s. The argument is unfortunately rather involved, owing principally to the lack of independence when considering the unrestricted process.

Here (and only here) the process d_{on} defined in §4.1 will be used.

<u>Lemma 5.9</u> $\lim_n \frac{d_{on}(\uparrow)}{n} = \lim_n \frac{d_{on}(\downarrow)}{n} = \mu(U)$ a.s.

<u>Proof</u>. We give the proof for $d_{on}(\uparrow)$ only, the other being identical. For ease of notation we drop the $\uparrow$ in the remainder of the proof.

First note that d_{mn} is a subadditive process; by Theorem 2.7 $\frac{d_{on}}{n}$ converges a.s. to a limit ξ. By considering first a "cylinder" version of d, it is established as in §5.1 that $\xi = \mu_d(U)$ a.s., where $\mu_d(U)$ is a constant. Because $a_{on}(\uparrow)$ is stochastically smaller than d_{on} it follows that $\mu(U) \leq \mu_d(U)$.

Consider next the processes d^k_{mn} defined in §4.1. For each k, this process is also subadditive and by arguing as above, there exists a constant $\mu^k(U)$ such that

(5.19) $\frac{d^k_{on}}{n} \to \mu^k(U)$ a.s., and $\mu(U) \leq \mu^k(U) \leq \mu_d(U)$.

For any fixed ω and n, d^k_{on} decreases as $k \to \infty$ to the first-passage time from (1,0) to (1,n); hence for n fixed, given any $\delta > 0$, there exists k(n) such that

(5.20) $E(\frac{d^k_{on}}{n}) \leq E(\frac{a_{on}}{n}) + \delta$ for $k > k(n)$.

Also, there exists $\tilde{n}$ such that

(5.21) $E(\frac{a_{o\tilde{n}}}{\tilde{n}}) \leq \mu(U) + \delta$.

Thus from (5.20) and (5.21),

(5.22) $\mu^k(U) \leq E(\frac{d^k_{o\tilde{n}}}{\tilde{n}}) \leq E(\frac{a_{o\tilde{n}}}{\tilde{n}}) + \delta \leq \mu(U) + 2\delta$ if $k > k(\tilde{n})$

and it follows from (5.19) that

(5.23) $\mu^k(U) \to \mu(U)$ as $k \to \infty$.

To complete the proof of Lemma 5.9 we show that $\mu^k(U) = \mu_d(U)$ for all k. Consider a route r from (1,0) to (1,n), lying in $\{x \geq 1-k\}$. Let r' be the path from (1 - k,0) to (1 - k,n) formed by first connecting (1 - k,0) and (1,0) by a straight line, then tracing the path r, then connecting (1,n) with (1 - k,n) by a straight line. Write $r' = r_1 * r * r_2$, where r_1 and r_2 are the paths over the horizontal lines from (1 - k,0) to (1,0) and from (1,n) to $(1-k,n)$ respectively. Since

$$t(r') = t(r_1) + t(r) + t(r_2)$$

and $t(r\ ) = d^k_{on}$, we have

$$(5.24) \qquad t(r') = \sum_{i=1}^{k} U_i^o + d^k_{on} + \sum_{i=1}^{k} U_i^n,$$

where U_i^m, $1 \leq i \leq k$, is the time coordinate of the bond from $(i-k,m)$ to $(i-k+1,m)$. Now let $d_{on}(1-k)$ be the first-passage time from $(1-k,0)$ to $(1-k,n)$ over paths in $\{x \geq 1-k\}$; evidently $d_{on}(1-k)$ has the same distribution as d_{on}. From (5.24),

$$(5.25) \qquad d_{on}(1-k) \leq \frac{t(r')}{n} \leq \frac{1}{n}\sum_{i=1}^{k} U_i^o + \frac{d^k_{on}}{n} + \frac{1}{n}\sum_{i=1}^{k} U_i^n.$$

The first term of the right-hand member of (5.25) clearly tends to zero as $n\to\infty$. Noting that $\sum_{i=1}^{k} U_i^n = S_n - S_{n-1}$, where $S_n = \sum_{m=1}^{n} \sum_{i=1}^{k} U_i^m$, it follows by the strong law of large numbers that $\frac{1}{n}\sum_{i=1}^{k} U_i^n \to 0$ a.s. as $n\to\infty$. Hence upon letting $n\to\infty$ in (5.25) we have

$$(5.26) \qquad \mu_d(U) \leq \mu^k(U)$$

which with (5.19) and (5.23) proves Lemma 5.9.

<u>Theorem 5.10</u> $\lim_n \frac{b_{on}}{n} = \mu(U)$ a.s.

<u>Proof</u>. If $\mu(U) = 0$ there is nothing to prove, since $0 \leq b_{on} \leq a_{on}$ a.s. and $\frac{a_{on}}{n} \to \mu(U)$ a.s. by Theorem 5.1. So we may assume $\mu(U) > 0$.

Fix a value of n and a route of b_{on}. In tracing out this path, starting at the origin, there will be a last point at which the path intersects the y-axis; let $k_n(\omega)$ denote the y-coordinate of this point. Let $s_{on}(k_n)$ denote the cylinder first-passage time from the point $(0,k_n)$ to the line $x = n$. Let r_1 denote the portion of the route for b_{on} from $(0,0)$ to $(0,k_n)$, and let r_2 denote the portion from $(0,k_n)$ to the line $x = n$. Since the chosen path is a route for b_{on}, it follows that there can be no faster path from $(0,0)$ to $(0,k_n)$; and r_2 must be the fastest cylinder path from k_n to the line $x = n$. Hence

$$(5.27) \qquad b_{on} = \begin{matrix} a_{ok_n}(\uparrow) + s_{on}(k_n) \\ a_{ok_n}(\downarrow) + s_{on}(k_n) \end{matrix}$$

according as k_n is positive or negative. Thus

$$(5.28) \qquad \frac{b_{on}}{n} = \frac{a_{ok_n}(\uparrow\downarrow)}{|k_n|} \cdot \frac{|k_n|}{n} + \frac{s_{on}(k_n)}{n} .$$

<u>Lemma 5.11</u> $\limsup_n \frac{|k_n(\omega)|}{n} \le 1$ a.s.

<u>Proof</u>. Suppose $\limsup_n \frac{k_n(\omega)}{n} = c > 1$ for some ω. Let $\{n_j\}$ be an increasing sequence of positive integers such that $k_{n_j}(\omega) \uparrow \infty$, $\lim_j \frac{k_{n_j}(\omega)}{n_j} = c$. From Theorem 5.1 we know that

$\frac{a_{ok_n}(\uparrow)}{k_n} \to \mu(U)$ a.s. along the subsequence $\{n_j\}$, so that from

(5.28),

$$\text{(5.29)} \qquad \limsup_n \frac{b_{on}(\omega)}{n} \geq \mu c > \mu .$$

But $\limsup_n \frac{b_{on}(\omega)}{n} \leq \lim_n \frac{a_{on}(\omega)}{n} = \mu$ a.s., so (5.29) occurs with zero probability; therefore $\limsup_n \frac{k_n(\omega)}{n} \leq 1$ a.s. Likewise $\liminf_n \frac{k_n(\omega)}{n} \geq -1$ a.s., proving the lemma.

Now let U_1 be the time coordinate of the bond joining $(0,0)$ to $(1,0)$. If $k_n(\omega) > 0$, let r' be the path which goes from $(0,0)$ to $(1,0)$, then follows a route of $d_{ok_n}(\uparrow)$ to $(1,k_n)$ (or until its intersection with r_2 defined above, whichever occurs first), then follows r_2 to the line $x = n$. If a route for $d_{ok_n}(\uparrow)$ hits the line $x = n$ before intersecting r_2 or hitting $(1,k_n)$, let r' end at the point where this route hits the line $x = n$. If $k_n(\omega) < 0$, r' is constructed analogously. It follows from Lemma 1.2 that

$$\text{(5.30)} \qquad t(r') \leq U_1 + d_{ok_n}(\uparrow\downarrow) + s_{on}(k_n) .$$

Using (5.27) and noting that r' is a cylinder path from the origin to the line $x = n$, we have

$$\text{(5.31)} \qquad b_{on} \leq s_{on} \leq b_{on} - a_{ok_n}(\uparrow\downarrow) + U_1 + d_{ok_n}(\uparrow\downarrow) .$$

Let Ω' be the event where all of the following occur:

(i) routes exist a.s. for b_{on}, $a_{on}(\uparrow)$, $d_{on}(\uparrow)$, $a_{on}(\downarrow)$, and $d_{on}(\downarrow)$ for all n;

(ii) $\lim_n \frac{d_{on}(\uparrow)}{n} = \lim_n \frac{d_{on}(\downarrow)}{n} = \mu(U)$;

(iii) $\lim_n \frac{a_{on}(\uparrow)}{n} = \lim_n \frac{a_{on}(\downarrow)}{n} = \mu(U)$;

(iv) $\limsup_n \frac{|k_n|}{n} \leq 1$;

(v) the time coordinates of all bonds are finite.

Fix an $\varepsilon > 0$ and an $\omega \in \Omega'$. There exists $n(\omega)$ such that for all $n \geq n(\omega)$:

(a) $U_1(\omega) < n\varepsilon$;

(b) $\left|\frac{a_{on}(\uparrow\downarrow)}{n} - \mu(U)\right| < \varepsilon$;

(c) $\left|\frac{d_{on}(\uparrow\downarrow)}{n} - \mu(U)\right| < \varepsilon$;

(d) $\left|\frac{k_n(\omega)}{n}\right| < 2$.

Also, there exists $M(\omega)$ such that $\max_{k<n(\omega)} \frac{d_{ok}(\uparrow\downarrow)}{m} < \varepsilon$ for $m \geq M(\omega)$. Now let

$$N(\omega) = \max\{n(\omega), M(\omega)\}.$$

Suppose $n \geq N(\omega)$. If $|k_n(\omega)| \geq n(\omega)$, then from (5.31):

(5.32) $$\frac{s_{on}}{n} \geq \frac{b_{on}}{n} \geq \frac{s_{on}}{n} - \frac{U_1}{n} + \left(\frac{a_{ok_n}^{(\uparrow\downarrow)}}{|k_n|} - \frac{d_{ok_n}^{(\uparrow\downarrow)}}{|k_n|}\right)\frac{|k_n|}{n}$$

$$\geq \frac{s_{on}}{n} - 5\varepsilon.$$

If $|k_n(\omega)| < n(\omega)$, then

(5.33) $$\frac{s_{on}}{n} \geq \frac{b_{on}}{n} \geq \frac{s_{on}}{n} - \frac{U_1}{n} - \frac{d_{ok_n}^{(\uparrow\downarrow)}}{n} \geq \frac{s_{on}}{n} - 2\varepsilon.$$

Therefore, on Ω', for every $\varepsilon > 0$,

$$\lim_n \frac{s_{on}}{n} \geq \limsup_n \frac{b_{on}}{n} \geq \liminf_n \frac{b_{on}}{n} \geq \lim_n \frac{s_{on}}{n} - 5\varepsilon .$$

Since $\frac{s_{on}}{n} \to \mu(U)$ a.s. by Corollary (5.6), and since ε is arbitrary, Theorem 5.10 is proved.

Comparison of Theorems 5.1 and 5.10 leads to the following interpretation: if fluid is introduced at each site along the line $x = n$, the time needed to wet the site $(0,0)$ is asymptotically the same as the time needed if fluid is introduced at the single site $(n,0)$.

The results of this chapter can be combined with the integrability results of §4.2 to give L^p-convergence of the first-passage processes.

> <u>Proposition 5.12</u> If the time coordinate distribution U has a finite m^{th} moment for $m \geq 1$, then
>
> $\frac{t_{on}}{n} \to \mu(U)$ in L^p for $p \leq m$;
>
> $\frac{s_{on}}{n} \to \mu(U)$ in L^p for $p \leq m$;
>
> $\frac{a_{on}}{n} \to \mu(U)$ in L^p for $p \leq 4m$;
>
> $\frac{b_{on}}{n} \to \mu(U)$ in L^p for $p \leq 4m$.

Proof. Theorems 5.1 and 5.10 give a.s. convergence of these processes; Theorem 4.5 gives uniform integrability of $\{\frac{t_{on}}{n}\}^p$ and $\{\frac{s_{on}}{n}\}^p$ for $p \leq m$ and Theorem 4.8 gives uniform integrability of $\{\frac{a_{on}}{n}\}^p$ and $\{\frac{b_{on}}{n}\}^p$ for $p \leq 4m$. The result follows in standard fashion (cf. [4], p.97).

5.4 Convergence of first-passage processes with negative time coordinates permitted

In §4.3 we proved the existence of routes for processes with time coordinate distribution $U \oplus r$, where $r < 0$, $U(0) < \lambda^{-1}$, and $\gamma(-r) < \lambda^{-1}$. We want now to consider the first-passage processes $t_{on}(\omega \oplus r)$, $s_{on}(\omega \oplus r)$, $a_{on}(\omega \oplus r)$, and $b_{on}(\omega \oplus r)$, where U and r satisfy the conditions above. The results presented below extend those of [50].

We note first that $t_{on}(\omega \oplus r)$ still satisfies conditions (2.5) and (2.6) for subadditive processes. For if $m < n < p$, a self-avoiding cylinder path from $(m,0)$ to $(n,0)$ will not intersect a self-avoiding cylinder path from $(n,0)$ to $(p,0)$; the connection of these paths is therefore a self-avoiding cylinder path from $(m,0)$ to $(p,0)$ and so

$$t_{mn}(\omega \oplus r) + t_{np}(\omega \oplus r) \geq t_{mp}(\omega \oplus r),$$

giving (2.5). Condition 2.6 is obvious since stationarity is preserved by adding the same constant to each time coordinate. However, (2.7) is no longer obvious since $t_{on}(\omega \oplus r)$ may be negative.

The unrestricted point-to-point time $a_{on}(\omega \oplus r)$ need not be subadditive. If the routes for $a_{mn}(\omega \oplus r)$ and $a_{np}(\omega \oplus r)$ intersect at a point other than $(n,0)$, let $\underset{\sim}{q}$ denote the first point, as the route of $a_{mn}(\omega \oplus r)$ is traveled from $(m,0)$, where the two routes intersect. If we connect the portion of this route before the intersection with the portion of the route of $a_{np}(\omega \oplus r)$ between $\underset{\sim}{q}$ and $(p,0)$, we get a self-avoiding path from $(m,0)$ to $(p,0)$; however, the deleted part may have negative travel time, so $a_{mn}(\omega \oplus r)$ need not satisfy (2.5).

Despite these difficulties it turns out, not surprisingly, that if the values of $r < 0$ are small enough, the first-passage processes all continue to converge a.s. to a common value. We begin with the point-to-point processes:

<u>Theorem 5.13</u> Suppose $r < 0$, $U(0) < \lambda^{-1}$, and $\gamma(-r) < \lambda^{-1}$. Then exists a non-negative constant $\mu(U \oplus r)$ such that

$$\lim_n \frac{t_{on}(\omega \oplus r)}{n} = \mu(U \oplus r) \text{ a.s.}$$

$$\lim_n \frac{a_{on}(\omega \oplus r)}{n} = \mu(U \oplus r) \text{ a.s.}$$

<u>Proof</u>. If $\underset{\sim}{m}, \underset{\sim}{n}$ are sites, let $a_{\underset{\sim}{m}\underset{\sim}{n}}$ denote the first-passage time from $\underset{\sim}{m}$ to $\underset{\sim}{n}$. Set

$$(5.34) \qquad v_n(\omega \oplus r) \equiv \sup_{\underset{\sim}{p} \in Z^2} \bar{a}_{(n,o)\underset{\sim}{p}}(\omega \oplus r)$$

where as usual x^- denotes the negative part of the function x. Then if $m<n<p$ (suppressing the $\omega\oplus r$),

(5.35) $$a_{mn}+a_{np}+2v_n \geq a_{mp}$$

so that

(5.36) $$a_{mn}+2v_n+a_{np}+2v_p \geq a_{mp}+2v_p$$

Now consider the non-negative process $\{a_{mn}(\omega\oplus r)+2v_n(\omega\oplus r)\}$. From (5.36) and the obvious stationarity, Theorem 2.7 can be applied provided that $v_n(\omega\oplus r)$ is integrable.

> Lemma 5.14 If $r<0$, $U(0)<\lambda^{-1}$, and $\gamma(-r)<\lambda^{-1}$, then $v_o(\omega\oplus r)$ has moments of all orders.

Proof. Given $\underset{\sim}{p}$, let $|\underset{\sim}{p}|$ denote the minimum number of bonds in a path from $(0,0)$ to $\underset{\sim}{p}$, i.e., the sum of the absolute values of the coordinates of $\underset{\sim}{p}$. For every $\underset{\sim}{p}$,

(5.37) $$P(a^-_{(0,0)\underset{\sim}{p}}(\omega\oplus r)>k) \leq \sum_{i\geq|\underset{\sim}{p}|} f_i P(S_i(\omega\oplus r)<-k),$$

where S_i is the sum of i independent random variables with the distribution U, and f_i, defined in §2.6, is the number of self-avoiding paths of i bonds leaving the origin. For every positive integer i,

(5.38) $$\begin{aligned} P(S_i(\omega\oplus r)<-k) &= P(S_i(\omega)<-k-ir) = P(S_i(\omega)+k+ir<0) \\ &= P(-u(S_i(\omega)+k+ir)>0) \quad \text{for any } u>0 \\ &\leq E(e^{-u}(S_i+k+ir)) = e^{-uk}E(e^{-u(S_i+ir)}). \end{aligned}$$

Choose $c < 1$ such that $\gamma(-r) < c^2\lambda^{-1}$. By definition of γ, there exists $u_o > 0$ such that $E(e^{-u_o(U_1+r)}) < c^2\lambda^{-1}$. Then

(5.39) $$P(S_i(\omega \oplus r) < -k) \leq e^{-u_o k}(c^2\lambda^{-1})^i.$$

By definition of λ (§2.5), there exists $A > 0$ such that

(5.40) $$f_i < A(\lambda c^{-1})^i \quad \text{for all } i.$$

So given any site $\underset{\sim}{p}$,

(5.41) $$P(a^-_{(0,0)\underset{\sim}{p}}(\omega \oplus r) > k) \leq \sum_{i \geq |\underset{\sim}{p}|} Ae^{-u_o k} c^i = Ae^{-u_o k} c^{|\underset{\sim}{p}|}(1-c)^{-1}$$

But for each positive integer n, there are exactly $4n$ points with $|\underset{\sim}{p}| = n$. Therefore,

(5.42) $$P(\sup_{\underset{\sim}{p} \in Z^2} a^-_{(0,0)\underset{\sim}{p}}(\omega \oplus r) > k) \leq \sum_{n=1}^{\infty} 4nAe^{-u_o k} c^n(1-c)^{-1}$$

$$= 4Ae^{-u_o k} c(1-c)^{-3}.$$

From (5.42) it follows that $\sum_{k=1}^{\infty} k^m P(v_o(\omega \oplus r) > k) < \infty$ for any positive m, proving Lemma 5.14.

Now applying Theorem 2.7 to the process $\{a_{mn}(\omega \oplus r) + 2v_n(\omega \oplus r)\}$,

(5.43) $$\lim_n \frac{a_{on}(\omega \oplus r) + 2v_n(\omega \oplus r)}{n} = \alpha_r(\omega) \text{ a.s. and in } L^1,$$

where $0 \leq \alpha_r(\omega) < \infty$. But since $E(v_o(\omega \oplus r) < \infty)$, for any $\epsilon > 0$ we have

(5.44) $$\sum_{n=1}^{\infty} P(\frac{v_n(\omega \oplus r)}{n} > \epsilon) = \sum_{n=1}^{\infty} P(v_o(\omega \oplus r) > n\epsilon) < \infty,$$

so by Borel-Cantelli, $\lim_n \frac{v_n(\omega \oplus r)}{n} = 0$ a.s. Thus

$$(5.45)\qquad \lim_n \frac{a_{on}(\omega\oplus r)}{n} = \alpha_r(\omega) \quad \text{a.s. and in } L^1.$$

But $a_{on}(\omega\oplus r) \leq t_{on}(\omega\oplus r)$ by definition so by Lemma 5.14,

$$(5.46)\qquad \sup E\left(\frac{t_{on}^-(\omega\oplus r)}{n}\right) \leq \sup_n E\left(\frac{a_{on}^-(\omega\oplus r)}{n}\right) < \infty.$$

By (5.46) and the remarks at the beginning of the section, $\{t_{mn}(\omega\oplus r)\}$ is an independent subadditive process and therefore (by Theorem 2.7):

$$(5.47)\qquad \lim_n \frac{t_{on}(\omega\oplus r)}{n} = \mu(U\oplus r) \text{ a.s. and in } L^1,$$

where $\mu(U\oplus r)$ is a constant depending only on the distribution U and on r.

It remains to be shown that $\alpha_r(\omega) = \mu(U\oplus r)$ a.s. Since $\alpha_r(\omega) \leq \mu(U\oplus r)$ a.s., it suffices to show that $E(\alpha_r) \geq \mu(U\oplus r)$. Let $q_{mn}^k(\omega\oplus r)$ be the first-passage time between $(m,0)$ and $(n,0)$ over all self-avoiding paths lying between the lines $x = m-k$ and $x = n+k$. Then for each k, $\{q_{mn}(\omega\oplus r) + 2v_n(\omega\oplus r)\}$ is a non-negative subadditive process and so converges a.s. and in L^1 to a random variable with expectation $\mu_k(U\oplus r)$. The proof may now be completed exactly as in Theorem 5.1, by showing that $\mu(U\oplus r) = \lim_k \mu_k(U\oplus r) = E(\alpha_r)$; then $\mu(U\oplus r)$ is non-negative and Theorem 5.13 is proved.

Finally we consider the point-to-line processes s and b. The a.s. convergence of these processes will be used only in §8.1 on length of routes since the complete proofs are technical and rather tedious, we give only a sketch of the arguments.

Theorem 5.15 Suppose $r < 0$, $U(0) < \lambda^{-1}$, $\gamma(-r) < \lambda^{-1}$. Then
$$\lim_n \frac{s_{on}(\omega\oplus r)}{n} = \lim_n \frac{b_{on}(\omega\oplus r)}{n} = \mu(U\oplus r) \text{ a.s.}$$

Proof. It follows just as in §5.2 that $\{s_{mn}(\omega\oplus r)\}$ is a process with subadditive expectations, that $z_{on}(\omega\oplus r)$ converges in probability, and that $\frac{z_{o,2^n m}(\omega\oplus r)}{2^n m}$ converges a.s. to the same limit for any positive integer m. However, $z_{on}(\omega\oplus r)$ may not be monotone in n so that a.s. convergence does not follow as before. The proof that $\frac{s_{on}(\omega\oplus r)}{n} \to \mu(U\oplus r)$ in probability and a.s. along any geometrically increasing subsequence can then be completed almost as in §5.2. The proof of Theorem 5.3 can be imitated in the present case, except for the use of barriers in Lemma 5.4 to establish that the height h_1 of the endpoint of the route for s_{on} has a finite second moment; with the possibility of negative time coordinates, the travel time over a portion of the path could exceed the travel time over the entire path, so the barrier argument no longer holds. In Chapter VIII we will prove this integrability result for $h_1(\omega\oplus r)$, using some route length results established there; let us therefore assume that $\frac{s_{on}(\omega\oplus r)}{n}$ has been shown to converge in probability to $\mu(U\oplus r)$, and a.s. along any geometrically increasing subsequence.

Although $s_{on}(\omega \oplus r)$ is not monotone, it is close enough to being monotone that the extension to a.s. convergence can be made in the usual way. It suffices to prove that

$$(5.48) \qquad P\Big(\inf_{2^n m < t \leq 2^n(m+1)} s_{ot} - s_{o,2^n m} < -\varepsilon 2^n m \quad \text{i.o. in } n\Big) = 0,$$

for any $\varepsilon > 0$ and any positive integer m. This is accomplished with the aid of Lemma 5.14 and a result from Chapter VIII on the route length for the process $s_{on}(\omega \oplus r)$ (Theorem 8.4).

Once the a.s. convergence of $s_{on}(\omega \oplus r)$ is established, the a.s. convergence of $b_{on}(\omega \oplus r)$ follows much as in Theorem 5.10, with a few essential modifications. The details of this argument may be found in [50] and will not be repeated here.

Finally we note that, in view of the exponential convergence of $P(\check{v}_o(\omega \oplus r) > k)$ (Lemma 5.14), the results of Theorems 4.5 and 4.8 on uniform integrability will continue to hold for the "shifted" processes, whenever $U(0) < \lambda^{-1}$ and $\gamma(-r) < \lambda^{-1}$.

Chapter VI

Renewal theory for percolation processes

Classical renewal theory (cf., for example, [11], Chapter XI) starts with an i.i.d. sequence $\{X_i\}_{i=1}^{\infty}$ of random variables, usually taken to be non-negative with a finite mean. If $S_n = X_1 + X_2 + \ldots + X_n$, and $t > 0$, $N(t)$ is defined to be the largest n such that $S_n \leq t$, and the renewal function $H(t)$ is defined to be $E(N(t))$.

In first-passage percolation, as in ordinary renewal theory, it is natural to ask how far (in a given direction) the fluid spreads in time t. Hammersley and Welsh (1965) considered this question for the cylinder processes and proved a "weak renewal theorem" in the case when the time coordinates are bounded above, and below away from zero. These results were extended to the unrestricted processes in [48]; in [49] and [60] the bounds were relaxed to a bound on the size of the atom at zero of the time coordinates. Also, in [48], [49], and [60], the a.s. convergence of the percolation analogues of $N(t)$ above was established. In section 6.1 we discuss these results; section 6.2 proves a weaker integrability result under a more general condition.

§6.1 Renewal theorems for the first-passage processes when $U(0) < \lambda^{-1}$

We begin by defining the "reach processes" which are the percolation analogues of the function $N(t)$.

Definition 6.1 For any $t \geq 0$,

$$x_t(\omega) = \sup\{n: t_{on}(\omega) \leq t\}$$

$$y_t(\omega) = \sup\{n: s_{on}(\omega) \leq t\}$$

$$x_t^u(\omega) = \sup\{n: a_{on}(\omega) \leq t\}$$

$$y_t^u(\omega) = \sup\{n: b_{on}(\omega) \leq t\}$$

Note that y_t^u is larger than x_t^u, x_t, and y_t, and that x_t is the smallest of these processes.

Lemma 6.2 For any time coordinate distribution U, $\lim_{t\to\infty} x_t = \infty$ a.s.

Proof. Let S_n denote the sum of the first n time coordinates along the x-axis (proceeding in the positive direction). Then if $N(t) = \sup\{n: S_n \leq t\}$, $N(t)\to\infty$ a.s., and $x_t \geq N(t)$.

Theorem 6.3 For any time coordinate distribution U, $\frac{x_t}{t} \xrightarrow{t\ \infty} [\mu(U)]^{-1}$ a.s. and similarly for x_t^u, y_t, and y_t^u.

Proof. If $P\{x_t(\omega) = \infty$ for some $t > 0\} > 0$, then $P(\liminf_n t_{on}(\omega) < \infty) > 0$, which by Theorem 5.1 implies that $\mu(U) = 0$; the theorem holds trivially in this case. Further, by Lemma 6.2, if $t_{ox_t} = 0$ for t arbitrarily large, then $\liminf_n t_{on} = 0$ and

$x_t = \infty$ for any $t > 0$; by the above analysis, the theorem holds in this case also. So we may assume that $x_t < \infty$ for all t and $t_{ox_t} > 0$ for t sufficiently large. Now

$$(6.1) \qquad \frac{x_t}{t} = \frac{x_t}{t_{ox_t}} \; \frac{t_{ox_t}}{t}$$

and $\frac{t_{ox_t}}{t} \to 1$ as $t \to \infty$ by definition of x_t and Lemma 6.2, so by Theorem 5.1, $\frac{x_t}{t} \to \mu^{-1}$ a.s. The proof for the other processes is identical (using Corollary 5.8 and Theorem 5.12 as needed).

A "weak renewal theorem" in this context would assert that $E(\frac{x_t}{t}) \to \mu^{-1}$, with a corresponding result for the other processes. To get a result like this we shall need to consider integrability of the reach process; for this we turn to an argument of Hammersley (1966).

> Definition 6.4 Let $B_n = \{(i,j) \in Z^2 : |i| + |j| = n\}$.
> Define $\eta_t \equiv \sup\{n: B_n$ can be reached by some path from $(0,0)$ in time $\leq t\}$.
> Let Z_n be the boundary of the square of side $2n$ and center at the origin.
> Define $z_t \equiv \sup\{n: Z_n$ can be reached by some path from $(0,0)$ in time $\leq t\}$.

Then $y_t^u \leq z_t \leq \eta_t$ so that moments of y_t^u (and thus a fortiori of the other reach processes) are dominated by the corresponding moments of η_t.

Lemma 6.5 Suppose $U(0) < \lambda^{-1}$, where U is the time coordinate distribution. Then

$$\sup_t E\left(\frac{\eta_t^k}{t^k}\right) \le C_k < \infty \quad \text{for} \quad k = 1,2,\ldots$$

Proof. Let p_k be the probability that there is a path of k bonds from $(0,0)$ with travel time less than Bk, where $B > 0$ By Lemma 2.12,

(6.2) $\qquad p_k \le f_k[\gamma(B)]^k.$

Since $\gamma(0) = U(0) < \lambda^{-1}$ and $\gamma(y)$ is right continuous by Theorem 2.11, we may choose $B > 0$ so small that for some $c < 1$,

(6.3) $\qquad \gamma(B) < c^2\lambda^{-1},$

and as in (5.40), there exists $A > 0$ such that

(6.4) $\qquad f_k < A(\lambda c^{-1})^k$ for all k.

By (6.2), (6.3), and (6.4):

(6.5) $\qquad p_k \le Ac^k$ for $k = 1,2,\ldots$

Then if $M_k(t) \equiv E(\eta_t^k)$,

(6.6) $$M_k(t) \le A \sum_{n=1}^{\infty} n^{k-1} P(\eta_t \ge n)$$

$$\le A''\left(\frac{t}{B}\right)^k + A' \sum_{n \ge \frac{t}{B}} n^{k-1} P(\eta_{nB} \ge n),$$

where the last inequality holds because $t \to \eta_t$ is monotonic increasing. But

(6.7) $\qquad P(\eta_{nB} \ge n) \le p_n,$

so by (6.5),

(6.8) $$M_k(t) \le A''(\tfrac{t}{B})^k + A\,A' \sum_{n=\frac{t}{B}}^{\infty} n^{k-1} c^n .$$

It follows from (6.8) that

(6.9) $$\frac{M_k(t)}{t^k} \le \frac{C_1}{B^k} + C_2,$$

where the right-hand side depends only on k and the distribution U.

Corollary 6.6 If $U(0) < \lambda^{-1}$, then $\{(\frac{y_t^u}{t})^p\}_{t\geq 0}$ is uniformly integrable for any $p > 0$ (and similarly for x_t, x_t^u, and y_t).

Combining Theorem 6.3 and Corollary 6.6, we have our "weak renewal theorem":

Theorem 6.7 If the time coordinate distribution satisfies $U(0) < \lambda^{-1}$, then
$\frac{y_t^u}{t} \xrightarrow{t\ \infty} \mu^{-1}$ in L^p for all $p > 0$;
in particular, $E(\frac{y_t^u}{t}) \xrightarrow{t\ \infty} \mu^{-1}$. Similar results hold for x_t, x_t^u, and y_t.

In Chapter VII we shall need a version of the weak renewal theorem with slightly different hypotheses. We begin with a lemma of Hammersley and Welsh ([27], p. 92):

Lemma 6.8 Let the time coordinates be bounded above by L and below by $B > 0$. Then
$E(y_{t_1}) + E(y_{t_2}) + 1 \leq E(y_{t_1+t_2+L})$.

Proof. Fix t_1; t_2 and let $y_{t_1}(\omega) = m_1$. Then by definition $s_{o,m_1}(\omega) \leq t_1$ and $s_{o,m_1+1}(\omega) > t_1$. Let r_1 be a route of $s_{o,m_1}(\omega)$ and let its endpoint be $P \equiv (m_1, z)$. Note that y_{t_1} depends only on the time coordinates of the bonds in the cylinder between $x = 0$ and $x = m_1 + 1$. Let ℓ be the bond from (m_1, z) to $(m_1 + 1, z)$ ($\equiv P'$). Since the time coordinate of ℓ does not exceed L, we have

(6.11) $t(r_1 * \ell, \omega) \leq t(r_1, \omega) + L$

Now let $y_{t_2}(P', \omega) \equiv \sup\{m: s_{om}(P') \leq t_2\}$, where $s_{om}(P')$ is the first-passage time from the point P' to the line $x = m_1 + 1 + m_2$. Suppose that $y_{t_2}(P', \omega) = m_2$; let r_2 be a path from P' to the line $x = m_1 + m_2 + 1$, such that

(6.12) $t(r_2, \omega) \leq t_2$.

The connected cylinder path $r_1 * \ell * r_2$ now links the origin to the line $x = m_1 + m_2 + 1$, and

(6.13) $t(r_1 * \ell * r_2) \leq t_1 + t_2 + L$.

Therefore,

(6.14) $\qquad y_{t_1+t_2+L}(\omega) \geq m_1 + m_2 + 1 = y_{t_1}(\omega) + y_{t_2}(P',\omega) + 1,$

and taking expectations in (6.14) gives

(6.15) $\qquad E(y_{t_1+t_2+L}) \geq E(y_{t_1}) + E(y_{t_2}(P',\omega)) + 1.$

But $y_{t_2}(P',\omega)$ is independent of P' and has the distribution, for any P', of y_{t_2}, giving us (6.10).

Corollary 6.9 The function $-1-E(y_{t-L})$ is subadditive in t, and is bounded below by $-1 - \frac{(t-L)}{B} \geq \frac{-t}{B}$.

With the aid of Corollary 6.9 we can prove the desired renewal theorem:

Theorem 6.10 Assume that the time coordinate distribution U is bounded above by L, and that $\mu(U) > 0$. Then

$$E\left(\frac{y_t}{t}\right) \xrightarrow{t\,\infty} \mu^{-1}.$$

Proof. Let U_B be the distribution obtained by truncating U below by B, i.e.

$$U_B(x) = 0 \quad \text{if} \quad x < B$$
$$= U(x) \quad \text{if} \quad x \geq B .$$

Let $\mu(U_B)$ be the time constant for this distribution, and y_t^B the corresponding reach process. Choose B to be less than the upper bound L of the distribution U. By Corollary 6.9 the function $E(y^B_{t-L}+1)$ is superadditive, and by Theorem 6.7, $E(\frac{y^B_t}{t}) \to [\mu(U_B)]^{-1}$; hence by (2.9) and (2.10),

$$\sup_t E\left(\frac{y^B_{t-L}+1}{t}\right) = [\mu(U_B)]^{-1}. \tag{6.16}$$

Thus for any t,

$$E\left(\frac{y^B_{t-L}+1}{t}\right) \le [\mu(U_B)]^{-1} \le [\mu(U)]^{-1} \quad \text{for all } B > 0. \tag{6.17}$$

But $y^B_t \le y_t$ for any B; define $\tilde{y}_t = \sup\{n: s_{on} < t\}$

and note that

$$y_t \ge \liminf_{B\downarrow 0} y^B_t \ge \tilde{y}_t, \tag{6.18}$$

so

$$E\left(\frac{\tilde{y}_t}{t}\right) \le E\left(\liminf_B \frac{y^B_t}{t}\right) \le \liminf_B E\left(\frac{y^B_t}{t}\right). \tag{6.19}$$

Given $\epsilon > 0$, use (6.17) to fix t_o such that

$$E\left(\frac{y^B_t}{t}\right) \le [\mu(U)]^{-1} + \epsilon \quad \text{for any } B > 0 \text{ and any } t > t_o \tag{6.20}$$

From (6.19),

$$E\left(\frac{\tilde{y}_t}{t}\right) \le [\mu(U)]^{-1} + \epsilon \qquad (t > t_o) \tag{6.21}$$

so

$$\limsup_{t\to\infty} E\left(\frac{\tilde{y}_t}{t}\right) \le [\mu(U)]^{-1}. \tag{6.22}$$

Noting finally that $\tilde{y}_t + 1 \ge y_t \ge \tilde{y}_t$, we get

$$\limsup_{t\to\infty} E\left(\frac{y_t}{t}\right) \le [\mu(U)]^{-1}. \tag{6.23}$$

Finally, $x_t \le y_t$ so by Theorem 6.3,

$$\liminf_t E(\frac{y_t}{t}) \ge \liminf_t E(\frac{x_t}{t}) \ge E(\lim_t \frac{x_t}{t}) = [\mu(U)]^{-1}. \tag{6.24}$$

From (6.23) and (6.24), the theorem is proved.

6.2 Integrability of the renewal process when $U(0) < p_T$.

Although it is not known if Corollary 6.6 can be extended to cases when $U(0) \ge \lambda^{-1}$, we can at least prove the integrability of the reach processes when $U(0) < p_T$, the "critical probability" defined in § 3.1.

Let η_t be as in Definition 6.4.

Theorem 6.11 If $U(0) < p_T$, $E(\eta_t) = o(t^{1+\delta})$ as $t \to \infty$ for any $\delta > 0$.

Proof. First assume the distribution U is Bernoulli. Given positive integers k and $n > k$,

$$\{\eta_k \ge n+k\} \subset \{\text{at least one bond in the diamond } \{|i|+|j| \le n+k\} \text{ is in an open cluster of size } \ge \frac{n}{k+1}\}$$

Hence for any positive $m \ge 1$,

$$P\{\eta_k \ge n+k\} \le 16n^2\, P\{(0,0) \text{ is in an open cluster of size } \ge \frac{n}{k+1}\}$$

$$\le 16n^2 \frac{E(|C|^m)}{(\frac{n}{k+1})^m} = \frac{16(k+1)^m E(|C|^m)}{n^{m-2}},$$

where we have used Lemma 3.11 to establish the finiteness of the mth

moment of $|C|$, the size of the open cluster containing $(0,0)$. Thus if $m \geq 4$,

$$E(\eta_k^{m-3}) \leq A \sum_{i=1}^{\infty} i^{m-4} P(\eta_k \geq i) \tag{6.25}$$

$$= A\{\sum_{i=1}^{2k} i^{m-4} P(\eta_k \geq i) + \sum_{i=2k+1}^{\infty} i^{m-4} P(\eta_k \geq i)\}$$

$$\leq \tilde{A}\{(2k)^{m-3} + \sum_{n=k+1}^{\infty} (n+k)^{m-4} P(\eta_k \geq n+k)\}$$

$$= A_m\{k^{m-3} + (k+1)^m \sum_{n=k+1}^{\infty} \frac{n^{m-4}}{n^{m-2}}\} = \tilde{A}_m (k+1)^m,$$

where $\tilde{A}_m$ does not depend on k. Taking roots in (6.25),

$$E(\eta_k) \leq \tilde{A}_m^{\frac{1}{m-3}} (k+1)^{\frac{m}{m-3}}$$

and it follows that $E(\eta_k) = o(k^{1+\delta})$ for any $\delta > 0$.

For a general distribution U with $U(0) < p_T$, we bound its reach process by the reach process of a Bernoulli distribution and apply the above result. Let $\epsilon > 0$ be chosen so small that $U(\epsilon) < p_T$ replace all time coordinates less than ϵ by zero and all those greater than ϵ by ϵ. The resulting distribution is Bernoulli (except for the scale factor ϵ) with atom at zero less than p_T, and its reach is clearly greater than the reach of U, for any time t. It follows from what has already been proved that

$$E(\eta_t) = o(t^{1+\delta}) \quad \text{as} \quad t\to\infty, \text{ for any } \delta > 0.$$

Chapter VII

The time constant

Up to this point we have had nothing to say about the value of the time constant $\mu(U)$, or its behavior as a functional of the distribution U. In this chapter we try to make amends, but unfortunately, the ergodic theorem for subadditive processes offers no help in evaluating μ; thus indirect means of doing so must be sought. Section 7.1 studies the problem of when $\mu(U)$ is zero. In §7.2 some upper bounds for μ are given, and in §7.3 we investi gate the behavior of μ as a functional of U.

7.1 Conditions for $\mu(U) = 0$.

One of the first questions that comes to mind about $\mu(U)$ is: For what distributions U is $\mu(U) = 0$? The results of the last chapter provide one bit of information about this problem:

> Proposition 7.1 If the time coordinate distribution U satisfies $U(0) < \lambda^{-1}$, then $\mu(U) > 0$.

Proof. From Corollary 6.6, $U(0) < \lambda^{-1}$ implies that the family $\{\frac{x_t}{t}\}$ is uniformly integrable, which means in particular that $\sup_t E(\frac{x_t}{t})$ is finite. But if $\mu(U)$ were zero, Theorem 6.3 would imply that $\frac{x_t}{t} \to \infty$ a.s. as $t \to \infty$; this cannot be, if the means are uniformly bounded. Hence $\mu(U) > 0$.

It is not known whether the converse of Proposition 7.1 is true.

Hammersley and Welsh conjectured in [27] that the condition $U(0) \geq \frac{1}{2}$ would imply that $\mu(U) = 0$. It was shown in [49] that $\mu(U) = 0$ whenever $U(0) > p_H$, where p_H is the critical probability defined in §3.1. The conjecture was verified completely in [61]; we state the result below in a more general form, in case p_T turns out to be less than one-half.

> Lemma 7.2 If $U(0) > p_T$ then $E(y_o) = \infty$, where y_o is the cylinder reach in zero time given in Definition 6.1.

Proof. The proof is much like that of Theorem 3.19. Let $U(0) = p$; regard a bond as being open if its time coordinate is zero, and closed otherwise. Recall from Chapter III (Theorem 3.8) that $p > p_T$ $(= p_S)$ implies that $\limsup_n S_n(p) > 0$, where $S_n(p)$ is the probability of a left-to-right crossing of the $n \times (n+1)$ sponge on open bonds, and p is the probability of a bond being open. Let $\{n_i\}$ be a subsequence such that $S_{n_i}(p) \geq \delta > 0$ for all i, for a fixed $p > p_T$. Let $T(n_i)$ denote the $n_i \times (n_i+1)$ sponge, and for $j = 0,1,\ldots,n_i - 1$, let $s_{on_i}(j, T(n_i))$ denote the first-passage time from the point $(0,j)$ to the line $x = n_i$ over paths which lie entirely in $T(n_i)$ and do not intersect the y-axis except at $(0,j)$. Since $S_{n_i}(p) \geq \delta$, it follows that

(7.1) $$P(s_{on_i}(j,T(n_i)) = 0) \geq \frac{\delta}{n_i} \quad \text{for some } j \in \{0,1,\ldots,n_i - 1\}.$$

Let $s_{on_i}(j)$ denote the cylinder first-passage time from $(0,j)$ to the line $x = n_i$; then it is clear from (7.1) that

$$P(s_{on_i}(j) = 0) \geq \frac{\delta}{n_i} \quad \text{for some } j \in \{0,1,\ldots,n_i - 1\}.$$

But $s_{on_i}(j)$ has the same distribution as s_{on_i}, so that

(7.2) $$P(s_{on_i} = 0) \geq \frac{\delta}{n_i}.$$

But $P(y_o \geq n_i) = P(s_{on_i} = 0)$ so from (7.2),

$$\limsup_n nP(y_o \geq n) \geq \delta > 0.$$

Hence y_o cannot be integrable.

Note that if $p = \frac{1}{2}$, then it follows from Theorem 2.2 that $S_n(\frac{1}{2}) = \frac{1}{2}$, and the above argument shows that $E(y_o) = \infty$ in this case as well.

<u>Theorem 7.3</u> If $U(0) > p_T$, then $\mu(U) = 0$.

If $U(0) = \frac{1}{2}$, then $\mu(U) = 0$.

<u>Proof</u>. The proof of both statements is the same. By Theorem 6.10, $E\left(\frac{y_t}{t}\right) \to [\mu(U)]^{-1}$ as $t \to \infty$, provided that $\mu(U) > 0$. But for any $t > 0$, $E(y_t) \geq E(y_o)$, so by Lemma 7.2 it follows that $\mu(U) = 0$

7.2 Bounds for $\mu(U)$

The determination of upper bounds for $\mu(U)$ has been pursued by Hammersley and Welsh (1965) and in [60]; to the best of our knowledge, nothing substantial is known concerning lower bounds for μ. The following theorem, due to Hammersley and Welsh, gives an obvious upper bound for $\mu(U)$ and shows that it is never attained (except trivially).

> Theorem 7.4 Let $\bar{u}$ be the mean of the distribution U. Then $0 \leq \mu(U) \leq \bar{u}$, and the second inequality is strict unless U is degenerate.

Proof. Let S_n be the sum of the first n time coordinates along the x-axis (traveling in the positive direction). Then

$$0 \leq \frac{t_{on}}{n} \leq \frac{S_n}{n}$$

and letting $n\to\infty$, by Theorem 4.5 and the strong law of large numbers we get $0 \leq \mu(U) \leq \bar{u}$. (Clearly if U is concentrated at one point $t_{on} = S_n$ and $\mu(U) = \bar{u}$.)

Now suppose U is not degenerate. Choose x such that $0 < x < \bar{u}$ and $U(x) > 0$; fix x and let $p = U(x)$. Choose n such that $(n+1)x < n\bar{u}$. Let $\ell_1, \ell_2, \ldots \ell_n$ be the first n bonds along the x-axis from $(0,0)$ to $(n,0)$. Let $\ell_1', \ell_2', \ldots, \ell_n'$ be the first n bonds along the line $y = 1$ from $(0,1)$ to $(n,1)$ and let ℓ_0' be the bond from $(0,0)$ to $(0,1)$. If U_i (resp. U_i') is the time coordinate of ℓ_i (resp. ℓ_i'), define

$$y(\omega) \equiv \min \left[\sum_{i=1}^{n} U_i, \sum_{i=0}^{n} U_i' \right].$$

Let

$$y^*(\omega) = \begin{cases} \sum_{i=0}^{n} U_i' & \text{if each } U_i' \leq x \\ \sum_{i=1}^{n} U_i & \text{otherwise} \end{cases}$$

then

$$E(y^*(\omega)) = p^{n+1} E\{ \sum_{i=0}^{n} U_i' \mid U_i' \leq x \text{ for all } i\} + (1-p^{n+1}) E[\sum_{i=1}^{n} U_i]$$

$$\leq p^{n+1}(n+1)x + (1-p^{n+1})n\bar{u} < n\bar{u}.$$

But clearly $s_{on}(\omega) \leq y(\omega) \leq y^*(\omega)$ and so

$$n\mu(U) \leq E(s_{on}) < n\bar{u},$$

proving Theorem 7.4.

A better upper bound, which can be calculated for some U, is given by Hammersley and Welsh ([27], p. 98). Let $Q(u) = 1-U(u)$. Let V be the cylinder $(0 \leq x \leq 1)$ first-passage time from the origin to the line $x = 1$, over paths lying strictly in the upper half-plane. Such a path proceeds by an upward step, followed by a step to the right or by another path of the same type from $(0,1)$, whichever is shorter; thus V is the independent sum of (1) the time coordinate of the bond from $(0,0)$ to $(0,1)$, and (2) the minimum of the time coordinate of the bond from $(0,1)$ to $(1,1)$ and a random variable with the distribution of V. Letting $G(x) = P(V > x)$, we have $1-G(x) = \int_0^x [1-Q(x-y)G(x-y)]dU(y)$, or

(7.3) $\qquad G(x) = Q(x) - \int_0^x Q(x-y)G(x-y)dQ(y)$.

Now let V_1 and V_2 be the cylinder $(0 \leq x < 1)$ first-passage times from the origin to the line $x = 1$ over paths lying strictly in the upper and lower half planes respectively, and let U_o be the time coordinate of the bond from $(0,0)$ to $(1,0)$. Let

(7.4) $\qquad g_{o1} \equiv \min (U_o, V_1, V_2)$.

Since V_1, V_2, and U_o are independent and $V_i (i=1,2)$ has the distribution of V above, it follows from (7.3) that

(7.5) $\qquad E(g_{o1}) \equiv \int_0^\infty Q(x) \, [G(x)]^2 dx$,

where G is the solution of (7.3).

Finally, if g_{mn} is defined to be the first-passage time between $(m,0)$ and the line $x = n$ over paths lying strictly between $x = m - 1$ and $x = n$, an argument exactly like that of Lemma 5.2 shows that the expectations $E(g_{on})$ form a subadditive function; since $b_{on} \leq g_{on} \leq s_{on}$, the time constant of the g-process is therefore equal to $\mu(U)$. By (2.9) and (2.10), and (7.5),

(7.6) $\qquad \mu(U) \leq E(g_{o1}) = \int_0^\infty Q(x)[G(x)]^2 dx$

where G is the solution of (7.3).

<u>Example 7.5</u> Let $Q(x) = e^{-x}$. Then (7.3) becomes

$$G(x) = e^{-x} + e^{-x} \int_0^x G(y)dy$$

so that G satisfies the differential equation

$$\frac{d}{dx} [e^x G(x)] = G(x),$$

which has the solution

$$G(x) = \exp(1-x-e^{-x}).$$

Plugging back into (7.6), this gives

(7.7) $\mu(U) \leq .59726.$

Example 7.6 Let U be the Bernoulli distribution with $P(U_i = 0) = p$. Straightforward calculation gives the following bounds:

$$p = \tfrac{1}{3} \qquad \mu(U) \leq .49$$

(7.8) $$p = \tfrac{1}{4} \qquad \mu(U) \leq .64$$

$$p = \tfrac{1}{10} \qquad \mu(U) \leq .85$$

A cruder method will yield an upper bound for the uniform (0,1) distribution. Let r_1 be the path from (0,0) to (0,1), then to (1,1), and let r_2 be its reflection in the x-axis; let r_3 be the one-bond path from (0,0) to (1,0). Then

$$g_{01} \leq \min\{t(r_1),\ t(r_2),\ t(r_3)\}$$

and $t(r_1)$, $t(r_2)$, and $t(r_3)$ are independent. Hence

$$\mu(U) \leq E(g_{01}) \leq E\,\min\{U_1 + U_2,\ U_3,\ U_4 + U_5\}$$

where U_i, $1 \leq i \leq 5$, are independent observations from U. For the uniform (0,1) distribution, this gives

(7.9) $\mu(U) \leq .425$

(Monte Carlo studies suggest that $\mu(U) \leq .328$ for this distributior (Welsh, 1965)).

For distributions with an atom at zero, [60] gives a bound for $\mu(U)$ in terms of the μ-value for a Bernoulli distribution with the same atom at zero.

<u>Theorem 7.7</u> Let $U(0) = p$, and let $\mu(p)$ be the time constant of the Bernoulli distribution with $P(U_i = 0) = p$. Then

$$\mu(U) \leq \frac{\mu(p)}{1-p}\,\bar{u}.$$

<u>Proof</u>. Let U be fixed, and define a distribution function F by

$$F(x) = 0 \qquad x \leq 0$$
$$= \frac{U(x)-p}{1-p} \qquad x > 0.$$

Then F is the distribution of a random variable from U conditioned on the event that it assumes positive values. Associated with the bonds $\{\ell_i\}$ of the lattice define an i.i.d. sequence $\{B_i\}$ of Bernoulli random variables with $p = P(B_1 = 0)$, and an independent sequence $\{X_i\}$ of i.i.d. random variables from F. Note that $B_i X_i$ has the distribution U.

Let r_n^B denote a route for s_{on}^B, the cylinder first-passage time with time states $\{B_i\}$. Let N_n denote the number of bonds in r_n^B which have $B_i = 1$. Then

$$s_{on}^B = N_n$$

and therefore

$$\lim_n E\left(\frac{N_n}{n}\right) = \mu(p). \tag{7.10}$$

Let t_n denote the travel time of the path r_n^B when the time states are $\{U_i\}$. Then t_n is the sum of the X_i on the N_n bonds of r_n^B with $B_i = 1$. The $\{X_i\}$ are independent of the $\{B_i\}$ and therefore of r_n^B, so that

(7.11) $\quad E(t_n \mid \{B_i\}) = E(X_1)\, N_n$

But $s_{on} \leq t_n$, so from (7.11):

$$E(\frac{s_{on}}{n}) \leq E(X_1)\; E(\frac{N_n}{n}).$$

Letting $n\to\infty$ and using (7.10),

$$\mu(U) \leq E(X_1)\; \mu(p).$$

But $E(X_1) = \frac{\bar{u}}{1-p}$, completing the proof.

7.3 The time constant as a function of the distribution U

It seems intuitively plausible (if not obvious) that the time constant $\mu(U)$ should be a reasonably smooth function of the underlying distribution U. In this section we present a few results in this direction.

7.3.1 Monotonicity of $\mu(U)$

The theorem below is obvious, and also true. It was established by Hammersley and Welsh ([27], p. 100):

> Theorem 7.8 If two underlying distribution functions U_1 and U_2 satisfy $U_1(u) \leq U_2(u)$ for all u, then $\mu(U_1) \geq \mu(U_2)$.

Proof. Given a distribution function $U(x)$, define its inverse in the usual (right-continuous) way:

(7.12) $U^{-1}(t) = \inf\{x: U(x) > t\} \qquad 0 \leq t < 1$

Evidently,

(7.13) $U_1^{-1}(t) \geq U_2^{-1}(t)$ if $U_1(u) \leq U_2(u)$ for all u.

By a well-known transformation, if ξ has a uniform $(0,1)$ distribution, then $U^{-1}(\xi)$ has the distribution U.

Now let the time coordinates $\{\xi_i\}$ be uniform $(0,1)$. Given ω, let $U^{-1}(\omega)$ be the sample point which assigns time coordinate $U^{-1}(\xi_i(\omega))$ to the bond ℓ_i whenever ω assigns $\xi_i(\omega)$ to ℓ_i. Then the time coordinates corresponding to $U^{-1}(\omega)$ have the distribution U, by the remark above. Thus given any ω and any path r,

(7.14) $t(r; U_1^{-1}(\omega)) \geq t(r; U_2^{-1}(\omega))$ if $U_1(u) \leq U_2(u)$ for all u.

Hence

(7.15) $t_{on}(U_1^{-1}(\omega)) \geq t_{on}(U_2^{-1}(\omega))$ for all ω.

Taking expectations in (7.15) and dividing by n, we get $\mu(U_1) \geq \mu(U_2)$ when $n \to \infty$.

Remark. The constant $\mu(U)$ is not a monotone function of the mean of U. To see this, let U_1 be a point mass at 0.45, and U_2 be uniform $(0,1)$. By (7.9),

$$\mu(U_2) \leq .425 \leq .45 = \mu(U_1),$$

but $\bar{u}_2 = .5 > .45 = \bar{u}_1$.

7.3.2 Concavity of $\mu(U\oplus r)$

Recall from Chapter IV the "shifted" time states $\omega\oplus r$, formed by adding the constant r to the time coordinate of each bond. Let q be any path, and $N(q)$ the number of bonds in the path. Then

$$t(q;\ \omega\oplus r) = t(q;\ \omega) + r\ N(q). \tag{7.16}$$

Therefore,

$$t_{on}(\omega\oplus r) \leq t_{on}(\omega) + r\ N_n(\omega), \tag{7.17}$$

where $N_n(\omega)$ is the number of bonds in the route of $t_{on}(\omega)$. Taking expected values in (7.17), we have

$$\tau_r(0,n) \equiv E(t_{on}(\omega\oplus r)) \leq \tau(0,n) + r\ E(N_n(\omega)). \tag{7.18}$$

The next theorem is a key result for Chapter VIII; it is a strengthened version of a result of Hammersley and Welsh ([27], p. 101).

> Theorem 7.9 Suppose $U(0) < \lambda^{-1}$. Let $D = R_+ \cup \{r<0:\ \gamma_U(-r) < \lambda^{-1}\}$. For $r\in D$, $\mu(U\oplus r)$ is a non-decreasing concave functional of r.

Proof. We can apply (7.18) to r and $-r$, assuming both are in D. Then

$$\tfrac{1}{2}[\tau_r(0,n) + \tau_{-r}(0,n)] \leq \tau(0,n) \tag{7.19}$$

By Theorems 5.1 and 5.13, we can divide by n and pass to the limit, giving

$$\tfrac{1}{2}\mu\ (U\oplus r) + \tfrac{1}{2}\mu\ (U\oplus -r) \leq \mu(U). \tag{7.20}$$

It follows easily from (7.20) that $\mu(U \oplus r)$ is a concave function of r; it is non-decreasing by Theorem 7.8.

Corollary 7.10 $\mu(U \oplus r)$ has left and right derivatives at each point in D (denoted by μ^- μ^+, respectively). μ^- is left continuous, μ^+ is right continuous, and $\mu^+ = \mu^-$ except possibly at countably many points.

$$\mu^-(r) \geq \mu^+(r) \geq 1 \quad \text{for all} \quad r \in D.$$

Proof. All but the last statement are standard facts about concave (convex) functions; see, for example, Saks (1964). For the last statement, it suffices to note that, from (7.18),

$$\tau_r(0,n) \leq \tau(0,n) + rn \quad \text{if} \quad r \leq 0,\ r \in D$$

because $N_n(\omega) \geq n$. Dividing by n and letting $n \to \infty$,

$$\mu(U \oplus r) \leq \mu(U) + r \quad \text{if} \quad r \leq 0,\ r \in D$$

and the result follows.

7.3.3 Truncation of U

Finally we consider the effect of truncation of U on the time constant $\mu(U)$. The two theorems presented below are taken from [49].

<u>Theorem 7.11</u> Let $\mu(U)$ be the time constant of the distribution U, let ${}^{B}U$ be the distribution obtained by truncating below at B, i.e.,

$$\begin{aligned} {}^{B}U(x) &= 0 \quad \text{if} \quad x < B \\ &= U(x) \quad \text{if} \quad x \geq B, \end{aligned}$$

and let $\mu({}^{B}U)$ be the time constant of this distribution. Then

$$\lim_{B\to 0} \mu({}^{B}U) = \mu(U).$$

<u>Proof</u>. Let $\{U_i\}$ have the distribution U, and let t_{on} be the cylinder first-passage time from the origin to (n,0) for the time state $\{U_i\}$. If $\tau(0,n) \equiv E(t_{on})$, then, given $\epsilon > 0$, there exists N_ϵ such that

$$n \geq N_\epsilon \Rightarrow \frac{\tau(0,n)}{n} - \mu(U) < \epsilon.$$

Fix $n = N_\epsilon$. If ${}^{B}t_{on}$ denotes the cylinder first-passage time from (0,0) to (n,0) for the time state $\{{}^{B}U_i\}$, where ${}^{B}U_i$ has the distribution ${}^{B}U$, then since routes exist a.s. for the processes $\{{}^{B}t_{on}\}$ and $\{t_{on}\}$

$$\lim_{B\to 0} {}^{B}t_{on} = t_{on} \quad \text{a.s.}$$

Hence if ${}^{B}\tau(0,m) \equiv E({}^{B}t_{on})$, it follows that there exists B such that

$$(7.21) \qquad B \leq B_\epsilon \Rightarrow \frac{{}^{B}\tau(0,n)}{n} - \mu(U) < 2\epsilon.$$

But according to Theorem 7.8, $\mu({}^{B}U)$ is a decreasing sequence as $B\downarrow 0$, so $\lim_{B\to 0} \mu({}^{B}U) \equiv \bar{\mu}$ exists; again by Theorem 7.8, $\mu({}^{B}U) \geq \mu(U)$ so that

(7.22) $\quad \bar{\mu} \geq \mu(U).$

To prove the opposite inequality, observe that since $\mu({}^{B}U) = \lim_{n} \frac{{}^{B}\tau(0,n)}{n} = \inf_{n} \frac{{}^{B}\tau(0,n)}{n}$, we have

(7.23) $\quad \bar{\mu} \leq \mu({}^{B}U) \leq \frac{{}^{B}\tau(0,n)}{n}$

and (7.21) and (7.23) together give

(7.24) $\quad \bar{\mu} - \mu(U) < 2\epsilon$, or $\bar{\mu} < \mu(U) + 2\epsilon$.

But $\epsilon > 0$ is arbitrary, so

(7.25) $\quad \bar{\mu} \leq \mu(U).$

From (7.22) and (7.25) we conclude that $\mu(U) = \bar{\mu} = \lim_{B\to 0} \mu({}^{B}U)$.

If we want to truncate off the right-hand tail, we have to work a little harder.

<u>Theorem 7.12</u> Let U be a distribution such that $U(a) = 0$ for some $a > 0$, and let $\mu(U)$ be the time constant of this distribution. Let U^B be the distribution obtained from U by truncating above at B, i.e.,

$$U^B(x) = U(x) \quad \text{if} \quad x < B,$$
$$= 1 \quad \text{if} \quad x \geq B,$$

and let $\mu(U^B)$ be the time constant of this distribution. Then $\lim_{B\to\infty} \mu(U^B) = \mu(U)$.

Proof. It is clear that $U(x) \leq U^B(x)$ for all x, so by Theorem 7.8,

(7.26) $\mu(U^B) \leq \mu(U)$.

Define a distribution function F^B by

$$F^B(x) \begin{array}{l} = 0 \quad \text{if} \quad x < 0 \\ = U(B) \quad \text{if} \quad 0 \leq x \leq B \\ = U(x) \quad \text{if} \quad x > B. \end{array}$$

Now let $\{U_i\}$ and $\{V_i\}$ be two i.i.d. sequences of random variables from the distribution U, independent of one another. For each $B > 0$, truncation of $\{U_i\}$ at B yields an i.i.d. sequence with the distribution U^B. Let 1_A denote the indicator function of the subset A of Ω. For given B, let $V_i^B \equiv V_i\, 1_{[V_i > B]}$; then the $\{V_i^B\}$ form an i.i.d. sequence from the distribution F^B defined above. Since the U_i^B and V_i^B are independent of each other for any B, the distribution function of $U_i^B + V_i^B$ is $U^B * F^B$, the convolution of V^B and F^B; it is straightforward to check that

$$U(x) \geq U^B * F^B(x) \quad \text{for any} \quad x.$$

By Theorem 7.8, it follows that

(7.27) $\mu(U) \leq \mu(U^B * F^B)$.

Let s_{on}^B denote the cylinder first-passage time from the origin to the line $x = n$ for the lattice with time state $\{U_i^B\}$, and let r_n^B denote a route for s_{on}^B. Now consider the lattice with time state $\{U_i^B + V_i^B\}$. The travel time of r_n^B for this distribution is $s_{on}^B + \sum_{\ell_i \in r_n^B} V_i^B$, where the sum is taken over all bonds in the route r_n^B. Since the family $\{V_i^B\}$ is independent of the family $\{U_i^B\}$ and the route r_n^B is completely determined by this latter family, the V_i^B are independent of r_n^B. Therefore,

$$(7.28)\qquad E\Big(\sum_{\ell_i \in r_n^B} V_i^B\Big) = E\Big(E\Big(\sum_{\ell_i \in r_n^B} V_i^B \mid \{U_i^B\}\Big)\Big) = E(N_n E(V_i^B))$$

$$= E(N_n)E(V_i^B),$$

where N_n denotes the number of bonds in the route r_n^B. Now since $E(V_i) < \infty$,

$$(7.29)\qquad \lim_{B\to\infty} E(V_i^B) = \lim_{B\to\infty} E(V_i\, 1_{[V_i > B]}) = 0.$$

Let s_{on}^+ denote the cylinder point-to-line time for the lattice with time state $\{U_i^B + V_i^B\}$. Clearly s_{on}^+ cannot exceed the travel time of r_n^B for this distribution, so from (7.28),

$$(7.30)\qquad E\Big(\frac{s_{on}^+}{n}\Big) \le E\Big(\frac{s_{on}^B}{n}\Big) + E\Big(\frac{N_n}{n}\Big)\, E(V_i^B).$$

Finally let s_{on} be the cylinder point-to-line time for the lattice with time state $\{U_i\}$. Since $U(a) = 0$,

$$N_n \le \frac{s_{on}^B}{a} \le \frac{s_{on}}{a},$$

so that

$$(7.31)\qquad E\Big(\frac{N_n}{n}\Big) \le E\Big(\frac{s_{on}^B}{an}\Big) \le E\Big(\frac{s_{on}}{an}\Big).$$

Letting $n\to\infty$ in (7.31) and using (7.30),

$$\lim_n E\Big(\frac{s_{on}^+}{n}\Big) \le \lim_n E\Big(\frac{s_{on}^B}{n}\Big) + \lim_n E(V_i^B)\, E\Big(\frac{s_{on}}{an}\Big),$$

so that (since $E(\frac{s_{on}}{n}) \to \mu(U)$),

(7.32) $$\mu(U^B * F^B) \le \mu(U^B) + \mu(U)\, E(\frac{V_i^B}{a}) .$$

But a is fixed; let $B\to\infty$ in (7.32). Then, using (7.26) and (7.27),

$$\limsup_{B\to\infty} \mu(U^B) \le \mu(U) \le \liminf_{B\to\infty} \mu(U^B * F^B)$$

$$\le \liminf_{B\to\infty} \mu(U^B) + \mu(U) \lim_{B\to\infty} \frac{E(V_i^B)}{a}.$$

By (7.29), we get finally that $\lim_{B\to\infty} \mu({}^BU) = \mu(U)$.

In Chapter VIII, after proving some results on route length, we will show how to replace the assumption $U(a) = 0$ in Theorem 7.12 by the condition $U(0) < \lambda^{-1}$. Finally we will show (with notation as in Theorem 7.7) that for Bernoulli distributions, $p\to\mu(p)$ is continuous for values of p in $[0,\lambda^{-1})$.

Chapter VIII

Route length and the height process

Two problems of obvious practical (as well as theoretical) interest will be considered in this chapter: given an optimal route for a first-passage process, how many bonds does it contain, and how high (or low) does it go?

The route length problem will be considered in sections 8.1 - 8.4, along with some numerical estimates. In the first three sections we look at time coordinate distributions with $U(0) < \lambda^{-1}$; in §8.4 the case $U(0) > p_H$ is treated. The techniques are completely different in the two cases; for values of $U(0)$ between λ^{-1} and p_H neither approach seems to work. Sections 8.5 - 8.7 are devoted to the height process, and in §8.8 the route length theorems are used to prove the two results mentioned at the end of Chapter VII.

8.1 Convergence a.s. of route lengths when $U(0) < \lambda^{-1}$

We first define the term "route length" as it is used in the first three sections:

<u>Definition 8.1</u> $N_n^t(\omega) \equiv \min \{k:$ there exists a route of $t_{on}(\omega)$ containing k bond

N_n^s, N_n^a, and N_n^b are defined analogously.

<u>Remark</u>. The definition given above ensures that N_n is well-defined. However, it will be clear from the proofs that follow in

sections 8.1 - 8.3 that the results of the theorems hold whenever N_n is the number of bonds in <u>any</u> route for the corresponding process.

A partial (asymptotic) answer to the route length problem was given by Hammersley and Welsh (1965) under the assumption that the time coordinates were bounded below away from zero. In this section we use the a.s. convergence results of Chapter V to prove a stronger result under the assumption that $U(0) < \lambda^{-1}$; most of what follows is taken (with some modification) from [50].

Recall from Corollary 7.10 that μ^+ and μ^- represent the right and left derivatives, respectively, of the function $r \to \mu(U \oplus r)$

<u>Theorem 8.2</u> Suppose that $U(0) < \lambda^{-1}$. Then

$$\mu^+(0) \leq \liminf_n \frac{N_n^{t,s,a,b}}{n} \leq \limsup_n \frac{N_n^{t,s,a,b}}{n} \leq \mu^-(0) \quad \text{a.s.}$$

<u>Proof</u>. We give the proof first for the t-process; the proof for the a-process is identical. From (7.17),

$$(8.1) \qquad t_{on}(\omega \oplus r) \leq t_{on}(\omega) + r\, N_n^t(\omega).$$

Fix $\epsilon > 0$ and $r > 0$. For almost all ω there exists $N(r,\omega)$ such that $n > N(r,\omega)$ implies

$$(8.2) \qquad \frac{t_{on}(\omega \oplus r)}{n} \geq \mu(U \oplus r) - \epsilon r$$

and

$$\frac{t_{on}(\omega)}{n} \leq \mu(U) + \epsilon r.$$

Combining (8.1) and (8.2),

$$(8.3) \qquad \mu(U\oplus r) - \mu(U) - 2\epsilon r \leq r \liminf_n \frac{N_n^t(\omega)}{n} \quad \text{a.s.}$$

Dividing by r and letting $r \to 0$ we get

$$(8.4) \qquad \mu^+(0) - 2\epsilon = \lim_{r\to 0} \frac{\mu(U\oplus r) - \mu(U)}{r} - 2\epsilon \leq \liminf_n \frac{N_n^t(\omega)}{n} \quad \text{a.s.}$$

Now take $r < 0$ so small that $\gamma(-r) < \lambda^{-1}$; for n sufficiently large we have from (8.1) that

$$(8.5) \qquad \frac{t_{on}(\omega\oplus r) - t_{on}(\omega)}{rn} \geq \frac{N_n^t(\omega)}{n} .$$

Applying Theorems 5.1 and 5.13,

$$(8.6) \qquad \frac{\mu(U\oplus r) - \mu(U)}{r} + 2\epsilon \geq \limsup_n \frac{N_n^t(\omega)}{n} \quad \text{a.s. for any } \epsilon > 0;$$

hence

$$(8.7) \qquad \mu^-(0) \geq \limsup_n \frac{N_n^t(\omega)}{n} \quad \text{a.s.}$$

Inequalities (8.4) and (8.7) together give the result for the processes t and a. For the s and b processes we proceed as above, using Corollary 5.8 and Theorems 5.10 and 5.15 instead of Theorems 5.1 and 5.13.

Of course, there is nothing sacred about the point $r = 0$:

<u>Corollary 8.3</u> If $U(0) < \lambda^{-1}$ and $r \leq 0$ satisfies $\gamma(-r) < \lambda^{-1}$, then

$$\mu^+(r) \leq \liminf_n \frac{N_n^{t,s,a,b}(\omega\oplus r)}{n}$$

$$\leq \limsup_n \frac{N_n^{t,s,a,b}(\omega\oplus r)}{n} \leq \mu^-(r). \text{ a.s.}$$

8.2 Numerical estimates for the route length

In view of the difficulty in determining the time constant μ, let alone its derivative at zero, it seems useful to take a different approach to get numerical bounds on $\frac{N_n}{n}$. (In this section and hereafter, we use the notation N_n to stand for any of $N_n^{\ t}$, $N_n^{\ s}$, $N_n^{\ a}$, $N_n^{\ b}$.) Two results are presented below; the second gives sharper bounds but is much less general. The first is taken from [50].

> Theorem 8.4 If $U(0) < \lambda^{-1}$, then there exists a constant k, depending only on U, such that $\limsup_n \frac{N_n}{n} \leq k$ a.s.

Proof. Take any $\rho > \mu(U)$, and choose $c < 1$ such that $U(0) < c^2\lambda^{-1}$. Then by Theorem 2.11, we can find $k > 0$ such that $\gamma(\frac{\rho}{k}) < c^2\lambda^{-1}$. With f_k and m_k defined as in §2.6,

$$(8.10) \qquad \sum_{n=1}^{\infty} P(m_{kn} < \rho n) \leq \sum_{n=1}^{\infty} f_{kn}\gamma(\tfrac{\rho}{k})^{kn} \leq \sum_{n=1}^{\infty} Ac^{kn} < \infty,$$

as in the proof of Lemma 5.14.

By the Borel-Cantelli lemma,

(8.11) $P(m_{kn} < \rho n \text{ i.o.}) = 0.$

Since $\frac{t_{on}}{n} \to \mu$ a.s.,

(8.12) $P(t_{on} > (\frac{\rho+\mu}{2})n \text{ i.o.}) = 0,$

and since t_{on} dominates the other first-passage times, (8.12) applies to b_{on}, s_{on}, and a_{on} as well. Therefore,

(8.13) $P\{\max\{t_{on}, s_{on}, a_{on}, b_{on}\} < m_{kn}$ for all n sufficiently large$) = 1$

and it follows that

(8.14) $P(N_n < kn$ for all n sufficiently large$) = 1,$

i.e., $\limsup_n \frac{N_n}{n} \leq k$ a.s.

In certain cases a bound on the constant k in Theorem 8.4 can easily be derived.

> Example 8.5 For the exponential distribution with parameter 1, $\gamma(y) = ye^{1-y}$ for $y < 1$.

Using the upper bound .59726 for μ give in (7.7), by the proof of Theorem 8.4 it is only necessary to choose k such that $\gamma(\frac{.60}{k}) < \lambda^{-1}$. Taking for λ^{-1} the approximate value .375, we find that $k \geq 3.7$ will work.

Example 8.6 For the uniform (0,1) distribution,

$$\gamma(y) = \inf_{u>0} e^{uy} \frac{(1-e^{-u})}{u}$$

and using the upper bound $\mu \leq .425$ given in (7.9), k need only be greater than 3.07.

Example 8.7 For the Bernoulli distribution with atom at zero of size p,

$$\gamma(y) = \left\{\frac{(1-p)(1-y)}{yp}\right\}^{y} \left(\frac{p}{1-y}\right) \quad \text{for} \quad y+p<1.$$

With $p = .25$, for example, using the upper bound $\mu \leq .64$ from (7.8), it is seen that $k \geq 6.9$ will suffice.

It is hardly necessary to point out that we are giving away a great deal in these bounds, since in all likelihood the true values of μ are much smaller than the bounds used above.

In certain cases a different approach gives sharper bounds than those in Examples 8.5 - 8.7. Consider a Bernoulli distribution U and let N_n^o denote the number of zero bonds in the route of (say) t_{on}. Then

(8.15) $$\frac{t_{on}}{n} = \frac{\text{number of one-bonds in the route}}{n} = \frac{N_n}{n} - \frac{N_n^o}{n} .$$

It follows from (8.15) that

(8.16) $$\limsup_n \frac{N_n(\omega)}{n} \leq \mu(U) + \limsup_n \frac{N_n^o(\omega)}{n} \quad \text{a.s.}$$

so that estimates on $\limsup_n \frac{N_n^o}{n}$, combined with bounds on μ, will yield bounds for $\frac{N_n}{n}$.

Pursuing this line of reasoning, let U be Bernoulli with $U(0)$ $(\equiv p) < \lambda^{-1}$. Let $\delta > 0$ be given. Then

(8.17) $P\{\exists$ a path with i zero-bonds and travel time $< n(\mu+\delta)\}$

$$\leq \sum_{k=i}^{k=i+n(\mu+\delta)} f_k p^i (1-p)^{k-i} \leq A(\lambda p)^i \sum_{i}^{i+n(\mu+\delta)} [\lambda(1-p)]^{k-i}$$

$$\leq A(\lambda p)^i \; [\lambda(1-p)]^{n(\mu+\delta)}$$

Given $\alpha > 0$, let E_n be the event $\{\exists$ a path with at least αn zero-bonds and travel time $< n(\mu+\delta)\}$. Then

(8.18) $$P(E_n) \leq A[\lambda(1-p)]^{n(\mu+\delta)} \sum_{i=\alpha n}^{\infty} (\lambda p)^i \sim (\lambda p)^{\alpha n}[\{\lambda(1-p)\}^{\mu+\delta}]^n$$

But

$$\{N_n \text{ has } \geq \alpha n \text{ zero-bonds}\} \subset E_n \cup \{t_{on} > n(\mu+\delta)\}$$

so that

$$\{N_n \text{ has } \geq \alpha n \text{ zero-bonds i.o.}\} \subset \{E_n \text{ i.o.}\} \cup \{t_{on} > n(\mu+\delta) \text{ i.o.}\}$$

Now $\{t_{on} > n(\mu+\delta)$ i.o.$\}$ is a null event, so by Borel-Cantelli, $\sum_n P(E_n) < \infty$ will imply that $P\{N_n$ has $\geq \alpha n$ zero-bonds i.o.$\} = 0$.

By (8.18),

(8.19) $$\sum_{n=1}^{\infty} P(E_n) \leq C \sum_{n=1}^{\infty} \{(\lambda p)^{\alpha}[\lambda(1-p)^{\mu+\delta}\}^n$$

and the sum on right-hand side of (8.19) converges provided that

$$(\lambda p)^{\alpha}[\lambda(1-p)]^{\mu+\delta} < 1 \quad \text{for some } \delta > 0.$$

Taking for λ the value 2.639, we then seek α_0 as small as possible and such that

(8.20) $(\lambda p)^{\alpha_o} [\lambda(1-p)]^{\mu} < 1.$

By (8.16), it then follows that

(8.21) $\limsup_n \frac{N_n(\omega)}{n} \leq \mu(U) + \alpha_o$ a.s.

Using the upper bounds for μ given in (7.8), inequality (8.21) gives improved upper bounds for $\limsup_n \frac{N_n}{n}$:

(8.22)

p	α_o	$\limsup \frac{N_n}{n}$
.33	1.9	2.39
.25	1.0	1.64
.10	.55	1.38

8.3 Integrability results for $\frac{N_n}{n}$ when $U(0) < \lambda^{-1}$

Recall the integrability results for the first-passage processes from §4.2, and the remark on integrability of the "shifted" processes at the end of Chapter V. These will be used in this section to study the integrability of $\frac{N_n}{n}$.

Theorem 8.8 Suppose that $U(0) < \lambda^{-1}$ and that $r \leq 0$ satisfies $\gamma(-r) < \lambda^{-1}$. If U has a finite m^{th} moment, where $m \geq 1$, then each of the families

$\{\frac{N_n^t(\omega\oplus r)}{n}\}^m$, $\{\frac{N_n^s(\omega\oplus r)}{n}\}^m$, $\{\frac{N_n^a(\omega\oplus r)}{n}\}^{4m}$, and

$\{\frac{N_n^b(\omega\oplus r)}{n}\}^{4m}$ is uniformly integrable.

Proof. We give the proof for $N_n^a(\omega\oplus r)$ only. Let $\bar{m}_k$ be the infimum of the travel times over all self-avoiding paths of k or more steps from $(0,0)$. Then for any $B > 0$,

(8.23) $$\{N_n^a(\omega\oplus r) \geq kn\} \subset \{\bar{m}_{kn}(\omega\oplus r) \leq a_{on}(\omega\oplus r)\}$$

$$\subset \{\bar{m}_{kn}(\omega\oplus r) \leq knB\} \cup \{a_{on}(\omega\oplus r) > knB\}$$

$$\subset \bigcup_{i=kn}^{\infty} \{m_i(\omega\oplus r) \leq iB\} \cup \{a_{on}(\omega\oplus r) > knB\}$$

Choose $c < 1$ such that $\gamma(-r) < c^2\lambda^{-1}$; choose $B < 0$ so small that $\gamma(B-r) < c^2\lambda^{-1}$. Let $A > 0$ be chosen so large that $f_i \leq A(\lambda c^{-1})^i$ for all i. Then

(8.24) $$P\{N_n^a(\omega\oplus r) \geq kn\} \leq \sum_{i=kn}^{\infty} f_i[\gamma(B-r)]^i + P\left(\frac{a_{on}(\omega\oplus r)}{n} > kB\right)$$

$$\leq Ac^{kn}(1-c)^{-1} + P\left(\frac{a_{on}(\omega\oplus r)}{n} > kB\right)$$

Hence

(8.25) $$E\left(\frac{N_n^a(\omega\oplus r)}{n}\right)^{4m} \leq 1 + A'(1-c)^{-1} \sum_{k=1}^{\infty} k^{4m-1} c^{kn}$$

$$+ A' \sum_{k=1}^{\infty} k^{4m-1} P\left(\frac{a_{on}(\omega\oplus r)}{nB} > k\right)$$

$$\leq 1 + A'(1-c)^{-1} \sum_{k=1}^{\infty} k^{4m-1} c^{kn} + A'B^{-4m} \sup_n E\left(\frac{a_{on}(\omega\oplus r)}{n}\right)^{4m}$$

Using Theorem 4.8 and the remark at the end of Chapter V, it follows that the expectations are uniformly bounded. To complete the proof of uniform integrability we show that:

(8.26) For every $\epsilon > 0$ there exists $\delta > 0$ such that $P(S) \leq \delta$ implies

$$\sup_n \int_S \left[\frac{N_n^{\,a}(\omega\oplus r)}{n}\right]^{4m} dP < \epsilon .$$

Now $S \cap \{N_n^{\,a}(\omega\oplus r) \geq kn\} \subset \bigcup_{i=kn}^{\infty} [S \cap \{m_i(\omega\oplus r) \leq iB\}]$

$\cup \{S \cap [\frac{a_{on}(\omega\oplus r)}{n} > kB]\}$, and by the Cauchy-Schwarz inequality,

$$(8.27)\qquad P(S \cap \{m_i(\omega\oplus r) \leq iB\}) = E\{1_S 1_{\{m_i(\omega\oplus r) \leq iB\}}\}$$

$$\leq [P(S)]^{\frac{1}{2}}\, P[m_i(\omega\oplus r) \leq iB]^{\frac{1}{2}} \leq [P(S)]^{\frac{1}{2}}\, A^{\frac{1}{2}}\, c^{\frac{i}{2}}$$

by Lemma 2.12 and the choice of A and c. Thus

$$(8.28)\qquad E\{1_S \left(\frac{N_n^a(\omega\oplus r)}{n}\right)^{4m}\} \leq P(S) + A' \sum_{k=1}^{\infty} \sum_{i=kn}^{\infty} k^{4m-1} [P(S)]^{\frac{1}{2}} A^{\frac{1}{2}} c^{\frac{i}{2}}$$

$$+ A' \sum_{k=1}^{\infty} k^{4m-1} P(S \cap \{\frac{a_{on}(\omega\oplus r)}{nB} \geq k\})$$

$$\leq P(S) + \tilde{A}\left\{[P(S)]^{\frac{1}{2}} + E\{1_S (\frac{a_{on}(\omega\oplus r)}{nB})\}^{4m}\right\}$$

where $\tilde{A}$ is a constant independent of n.

By the uniform integrability of $\{\frac{a_{on}(\omega\oplus r)}{n}\}^{4m}$, the last term of (8.28) may be made less than $\frac{\epsilon}{2}$ by making $P(S)$ less than some $\delta > 0$, so that (8.26) holds and the theorem is proved.

Combining Theorem 8.8 with Theorem 8.2 and Corollary 8.3, we have the following result on L_p convergence:

Theorem 8.9 Suppose $U(0) < \lambda^{-1}$ and $r \leq 0$ satisfies $\gamma(-r) < \lambda^{-1}$. Let U have a finite m^{th} moment, where $m \geq 1$. Then, provided that $\mu^{+}(r) = \mu^{-}(r)$ $(= \mu'(r)$, say$)$,

$$\lim_n \frac{N_n^t(\omega\oplus r)}{n} = \lim_n \frac{N_n^s(\omega\oplus r)}{n} = \mu'(r) \quad \text{in } L_p \text{ for}$$

$0 < p \leq m$;

$$\lim_n \frac{N_n^a(\omega\oplus r)}{n} = \lim_n \frac{N_n^b(\omega\oplus r)}{n} = \mu'(r) \quad \text{in } L_p \text{ for}$$

$0 < p \leq 4m$.

8.4 Route length bounds when $U(0) > p_H$

The results of Chapter III will be used in this section to describe, for sufficiently large n, how to construct a route which contains $O(n)$ bonds. From Definition 8.1 it will then follow that $\limsup_n \frac{N_n}{n} \leq k$ a.s. for some constant k.

Theorem 8.10 Suppose $U(0) > p_H$. Let $|C|$ denote the number of bonds in the closed cluster containing a given bond in a Bernoulli percolation model with $p = U(0)$. Then

$$\limsup_n \frac{N_n(\omega)}{n} \leq E|C| + 2 - U(0) \text{ a.s.}$$

Proof. We first give the proof for N_n^b, with modifications sketched later for the other first-passage processes. Consider a bond to be open if its time coordinate is zero and closed otherwise. Let $p = U(0)$, $q = 1 - p$; note that by Theorem 3.8, $p > p_H$ implies $q < p_T$, so the expected size of the closed cluster containing a given bond is finite.

Since $p > p_H$, there is an event $\tilde{\Omega}$ of probability one on which there exists an infinite cluster of open bonds, and by Lemma 3.6 there are an infinite number of disjoint open circuits surrounding the origin. Given $\omega \in \tilde{\Omega}$, we choose N so large that the square A_N of side N, centered at the origin, contains an open circuit which is in the infinite cluster of open bonds.

There are now two possibilities: either there exists a closed circuit in the dual lattice L^* with $(0,0)$ in its interior, or there does not. Since the path construction for the second case will be seen to follow from the first case, we consider the first case. We look at all closed circuits in L^* around $(0,0)$, excluding any which are completely surrounded by another one. The union of all remaining closed circuits is a connected set of bonds. Let G_o^* denote the closed cluster in L^* which contains this union; then G_o^* is contained in A_N, and G_o^* is surrounded by a cut set of open bonds in L^*. By Whitney's Theorem (Theorem 2.1), there is a minimal open circuit Δ_o in L which surrounds G_o^* and thus also the origin. For n sufficiently large, no point on the line $x = n$ lies within Δ_o; we consider only such values of n. For any k

with $0 \leq k < n$, if the bond in L^* connecting $(k+\frac{1}{2}, \frac{1}{2})$ with $(k+\frac{1}{2}, -\frac{1}{2})$ is closed, let C_k^* denote the closed cluster in L^* containing it. This closed cluster in L^* is also surrounded by a minimal open circuit D_k in L containing $(k+\frac{1}{2}, 0)$ in its interior.

A route for $b_{on}(\omega)$ is then found as follows. First, we find a path from the origin to the circuit Δ_o which has the shortest possible travel time (of course, such a path need not be unique). Now the circuit Δ_o, by construction, is part of an infinite open cluster; the rest of the route of $b_{on}(\omega)$ will be chosen through bonds lying in this cluster. To do this, proceed along the circuit Δ_o to its intersection with the positive x-axis; if there is more than one such intersection, proceed to the one with the greatest x-coordinate. Denote by $X_o(\omega)$ the number of bonds in the path from (0,0) out to Δ_o and on to this intersection with the x-axis. Now proceed to the right along the x-axis as far as possible along open bonds. When a closed bond is encountered, the leftmost vertex (k,0) of this bond must belong to an open cluster containing the circuit D_k in L. Travel the circuit D_k (in the direction giving the fewest bonds) until the x-axis is intersected again at (k',0) where $k' > k$. (To do this we may have to backtrack initially along the x-axis from (k,0), so the path may be self-intersecting; once the path to the line $x = n$ is constructed we can then eliminate all such loops. This can only decrease the number of bonds in the path, and ensures that each open bond along the x-axis is travelled at most once.) Now repeat the process above; travel the x-axis along open bonds until a closed bond is encountered, at which point the corresponding open circuit D_j is travelled back to the x-axis at some $j' > j$. In this way the route must eventually reach the line $x = n$. (Figure 2 sketches a possible such route for b_{on}).

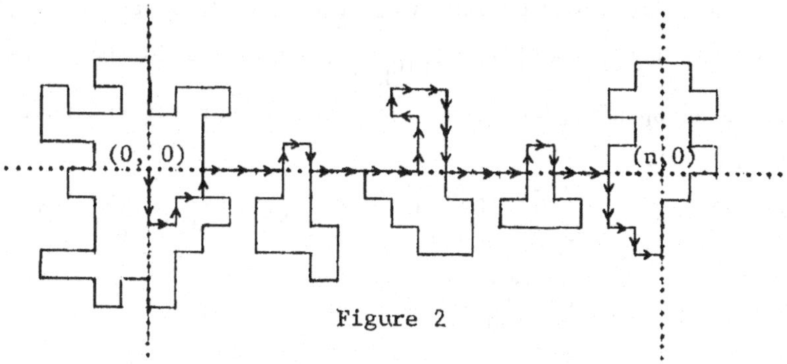

Figure 2

Consider now the second case outlined above. If there is no closed circuit in L^* with $(0,0)$ in its interior, then $(0,0)$ belongs to an infinite open cluster. In this case we simply start off from $(0,0)$ by travelling the x-axis along zero bonds as far as possible and repeating the process described above.

We next find an upper bound for the number of bonds in the path just constructed. First, $|D_k|$ is at most $2|C_k^*| + 4\, 1_{\{|C_k^*| > 0\}}$.

To see this, by induction, note that six bonds in L are needed to enclose one bond in L^*; adding an additional bond to a closed cluster requires at most two additional bonds in the enclosing circuit, regardless of the cluster size or shape. (The worst that can happen is that we may add three bonds and erase one). For each D_k we chose for our route the portion with the fewest bonds between the corresponding intersections with the x-axis, so at most half the bonds of D_k, i.e., $|C_k^*| + 2\, 1_{[|C_k^*|>0]}$, belong to the route.

Since an open bond on the x-axis may be part of the route, we have finally that

$$(8.29) \qquad N_n^b \le X_o + \sum_{k=0}^{n-1} (|C_k^*| + 2\, 1_{[|C_k^*|>0]} + 1_{[|C_k^*|=0]})$$

$$= X_o + n + \sum_{k=0}^{n-1} (|C_k^*| + 1_{[|C_k^*|>0]})$$

The variables $\{|C_k^*| + 1_{[|C_k^*|>0]}\}$ form a stationary sequence, and

$$E(|C_k^*| + 1_{[|C_k^*|>0]}) = E|C| + q < \infty$$

by the remarks at the beginning of the proof.

Lemma 8.11 The sequence $\{|C_k^*| + 1_{[|C_k^*|>0]}\}$ is ergodic.

Proof. If ℓ is any configuration of open and closed bonds in L^*, let τ be the transformation that shifts ℓ one unit to the right. If A is an event, then τA is the event obtained by shifting each $\ell \in A$ one unit to the right. τ is measure-preserving ($P(A) = P(\tau A)$ for any event A) and of mixing type, i.e., $\lim_n P(A \cap \tau^n B) = P(A)\,P(B)$ for any events A and B. This is clear when A and B are cylinder sets, since then A and $\tau^n B$ are independent for n sufficiently large. For a general event C, given $\epsilon > 0$ we can find a cylinder set D such that $P(C \Delta D) < \epsilon$ (cf. [4], p. 251); the general case follows from this fact. Since τ is of mixing type, it follows easily that τ is ergodic (i.e., metrically transitive).

By Lemma 8.11 and the Birkhoff ergodic theorem it follows that

$$(8.30) \qquad \lim \frac{1}{n} \sum_{k=0}^{n-1} \left(|C_k^*| + 1_{[\,|C_k^*| > 0]}\right) = E|C| + q \quad \text{a.s.}$$

Clearly $\frac{X_o}{n} \to 0$ a.s. as $n \to \infty$, so from (8.29) and (8.30) it follows that

$$\limsup_n \frac{N_n^{b}(\omega)}{n} \leq E|C| + 1 + q = E|C| + 2 - U(0) \quad \text{a.s.},$$

proving Theorem 8.10 for the b-process.

To modify the proof above for the a-process, the following lemma is needed:

<u>Lemma 8.12</u> The random variable X_o has moments of all orders.

<u>Proof</u>. For some $j > 0$, the closed cluster G_o^* in the dual contains the bond joining $(j + \frac{1}{2}, \frac{1}{2})$ with $(j + \frac{1}{2}, -\frac{1}{2})$. Let j_o denote the minimal such j. Then G_o^* is identical with $C_{j_o}^*$, so

$$(8.31) \qquad E(|G_o^*|^j) = \sum_{k=1}^{\infty} E[|C_k^*|^j 1_{[j_o = k]}]$$

$$\leq \sum_{k=1}^{\infty} [E|C_k^*|^{2j}]^{\frac{1}{2}} \sqrt{P(j_o = k)} \leq \sqrt{\alpha_{2j}} \sum_{k=1}^{\infty} \sqrt{P(j_o = k)} \ ,$$

where $\alpha_{2j} \equiv E\ |C|^{2j}$ is finite by the argument given in Lemma 3.10. But

$$P(j_o = k) \leq P(|C_k^*| > k) \leq \frac{E(|C|^4)}{k^4} \leq \frac{\alpha_4}{k^4} \ ,$$

so $\sqrt{P(j_o = k)} \leq \frac{\sqrt{\alpha_4}}{k^2}$ and by (8.31), $E(|G_o^*|^j) < \infty$ for any $j \geq 1$. Since X_o cannot exceed a constant times $|G_o^*|^2$, the lemma is proved.

To prove Theorem 8.10 for the a-process, construct the open circuit Δ_o around $(0,0)$ as before; given a positive integer n, construct the closed cluster G^*_{n-1} in L^* around $(n-\frac{1}{2}, \frac{1}{2})$ the same way G_o^* was constructed around $(0,0)$ (if there is any such closed cluster). Surround G^*_{n-1} with an open circuit Δ_{n-1} containing $(n,0)$ in its interior; for n sufficiently large, Δ_{n-1} and Δ_o will be disjoint. Let X_{n-1} denote the number of bonds in the quickest path from $(n,0)$ to Δ_{n-1} and back to the x-axis. Then construct the path for $a_{on}(\omega)$ just as for $b_{on}(\omega)$, except that once Δ_{n-1} is encountered, we follow it to its intersection with the path to $(n,0)$. (If no closed cluster exists surrounding $(n-\frac{1}{2}, \frac{1}{2})$, we either arrive at $(n,0)$ along the x-axis or else some open circuit D_k leads to $(n,0)$).

This gives a route for $a_{on}(\omega)$, and the analogue of (8.29) is now

(8.32) $$N_n^a(\omega) \leq X_o + X_{n-1} + n + \sum_{k=0}^{n-1} (|C_k^*| + 1_{\{|C_k^*|>0\}}).$$

By Lemma 8.12 and Borel-Cantelli, $\frac{X_{n-1}}{n} \to 0$ a.s. as $n \to \infty$, so we get

$$\limsup_n \frac{N_n^a(\omega)}{n} \leq E|C| + 2 - U(0)$$

as before.

Finally we consider the cylinder processes s and t. The route construction algorithm used above for b does not necessarily produce a cylinder route since the open circuits D_k may intersect $\{x \leq 0\}$. Note, however, that by Theorem 3.8, $p_H^+ = p_H$. Thus with probability one there exists an infinite open cluster contained in the quadrant $\{x \geq 1, y \geq 0\}$, and an infinite open cluster contained in the quadrant $\{x \geq 1, y \leq 0\}$. We now choose N so large that the square A_N of side N centered at (1,0) contains an open half circuit in $\{x \geq 1\}$ around (1,0) which intersects both of these infinite open clusters; G_o^* will then be a closed cluster in $L^* \cap \{x \geq \frac{3}{2}\}$ which is contained in A_N and surrounds (1,0), and Δ_o will be an open half circuit in $L \cap \{x \geq 1\}$ which surrounds (1,0). The route for $s_{on}(\omega)$ begins with a step along the x-axis and then travels the quickest path to Δ_o; the rest of the path proceeds as before and is necessarily a cylinder path since Δ_o now intersects the two infinite clusters described above. If there exists no G_o^*

we simply move along the x-axis until the first closed bond is encountered. A route for t_{on} is constructed similarly, making the cylinder modification at $(n,0)$ corresponding to the modification made for a_{on}. We finally arrive at

$$\limsup_n \frac{N_n^{s,t}(\omega)}{n} \leq E|C| + 2 - U(0) \quad \text{a.s.},$$

completing the proof of Theorem 8.10.

<u>Corollary 8.13</u> $\limsup_n E(\frac{N_n}{n}) \leq E|C| + 2 - U(0)$.

<u>Proof</u>. We can take expectations on both sides of the inequalities (8.29) and (8.32) and their cylinder counterparts, by Lemma 8.12; since the Birkhoff ergodic theorem gives convergence in L^1 as well as a.s., there is no problem in passing to the limit.

8.5 Bounds for the height process when $U(0) < \lambda^{-1}$

The "height process" is defined as follows:

<u>Definition 8.14</u> If r is a route for t_{on}, let

$h_n^t(r,\omega) \equiv \max\{|m|: (k,m)$ is a site on r for some $k\}$

$\bar{h}_n^t(\omega) \equiv \max\{h_n^t(r,\omega): r$ is a route for $t_{on}\}$

$\underline{h}_n^t(\omega) \equiv \min\{h_n^t(r,\omega): r$ is a route for $t_{on}\}$.

The height processes for s, a, and b are defined similarly.

It should be noted that the "height problem" mentioned by Hammersley and Welsh ([27], p. 108-09) considers a different "height", i.e., the y-coordinate of the terminal site of a route for s_{on} or b_{on}. Clearly the absolute value of this height is not greater than $h_n^{s,b}(r,\omega)$ above.

In this section we note some implications for the height process of preceding results. As before, $\bar{h}_n$ or $\underline{h}_n$, without an identifying superscript, denotes any of the four height processes.

<u>Theorem 8.15</u> Let U be a time coordinate distribution with $\mu(U) > 0$. Then
$$\limsup_n \frac{\bar{h}_n(\omega)}{n} \leq 1 \quad \text{a.s.}$$

<u>Proof</u>. Suppose $P\{\bar{h}_n(\omega) > n(1+\delta) \text{ i.o.}\}$ is positive. Then (in the notation of §4.1),

Either $b_{o,n(1+\delta)}(\uparrow) \leq b_{on}$ i.o.

or $b_{o,n(1+\delta)}(\downarrow) \leq b_{on}$ i.o.

with positive probability.

Assuming the first case,
$$\frac{b_{o,n(1+\delta)}(\uparrow)}{n(1+\delta)} \leq \frac{b_{on}}{n}\left(\frac{1}{1+\delta}\right) \quad \text{i.o. with positive probability.}$$

By Theorem 5.10, and noting that $b_{on}(\uparrow)$ has the same distribution as b_{on}, we get
$$\mu \leq \mu(1+\delta)^{-1},$$

a contradiction. Hence $\limsup_n \frac{\bar{h}_n(\omega)}{n} \leq 1+\delta$ a.s., and since δ is arbitrary, the result follows.

Corollary 8.16 If $U(0) < \lambda^{-1}$, then
$$\limsup_n \frac{\bar{h}_n(\omega)}{n} \leq 1 \quad \text{a.s.}$$

Proof. This is immediate from Proposition 7.1 and Theorem 8.15.

Recall the remark following Definition 8.1; since $h_n(\omega)$ can never exceed {the number of bonds in r} - n (or one-half this number for the processes a_{on} and t_{on}), the results of §8.2 allow some improvement on Theorem 8.15 in particular cases. For example, it follows from (8.22) that for Bernoulli distributions with atoms at zero of size .25 and .10, $\limsup_n \frac{\bar{h}_n(\omega)}{n} \leq .64$ and .38, respectively, for the s and b processes, and .32 and .19 for the a and t processes. From Corollary 8.3 we can deduce a general result:

Corollary 8.17 If $U(0) < \lambda^{-1}$ and $r \leq 0$ satisfies $\gamma(-r) < \lambda^{-1}$, then
$$\limsup_n \frac{\bar{h}_n^{s,b}(\omega \oplus r)}{n} \leq \mu^-(r) - 1 \quad \text{a.s.}$$
$$\limsup_n \frac{\bar{h}_n^{a,t}(\omega \oplus r)}{n} \leq \tfrac{1}{2}(\mu^-(r) - 1) \quad \text{a.s.}$$

To close this section we prove the result about the height process which was needed to establish convergence in probability of $\frac{s_{on}(\omega\oplus r)}{n}$ in Theorem 5.15.

Theorem 8.18 Suppose that $U(0) < \lambda^{-1}$, $r \leq 0$ satisfies $\gamma(-r) < \lambda^{-1}$, and that U has a finite m^{th} moment for some $m \geq 1$. Then each of the families

$$\left\{\frac{\underline{h}_n^{\,t}(\omega\oplus r)}{n}\right\}^{4m}, \quad \left\{\frac{\underline{h}_n^{\,s}(\omega\oplus r)}{n}\right\}^{4m}, \quad \left\{\frac{\bar{h}_n^{\,a}(\omega\oplus r)}{n}\right\}^{4m},$$

$$\left\{\frac{\bar{h}_n^{\,b}(\omega\oplus r)}{n}\right\}^{4m}$$

is uniformly integrable.

Proof. For a and b this is immediate from Theorem 8.8; for s and t a special argument is needed.

Let b_{on}^{+} (resp., b_{on}^{-}) denote the first-passage time from $(0,2)$ (resp., $(0,-2)$) to the line $x = n$ over paths contained entirely in the half-plane $y \geq 1$ (resp., $y \leq -1$). It follows as for b_{on} that routes exist a.s. for $b_{on}^{\pm}(\omega\oplus r)$, and the family $\left\{\frac{b_{on}^{\pm}(\omega\oplus r)}{n}\right\}^{4m}$ is uniformly integrable. Thus by the argument of Theorem 8.8, the route length families $\left\{\frac{N_n^{b^{\pm}}(\omega\oplus r)}{n}\right\}^{4m}$ are uniformly integrable.

Let $P_n^{\pm}$ denote the sites at which some routes for $b_{on}^{\pm}(\omega\oplus r)$ last intersect the y-axis; it is easily seen that there must be routes for $s_{on}(\omega\oplus r)$ and for $t_{on}(\omega\oplus r)$ which do not cross the portions of the routes for $b_{on}^{\pm}(\omega\oplus r)$ lying between $P_n^{\pm}$ and the line $x = n$. Therefore,

(8.33) $\underline{h}_n^{s,t} \leq N_n^{b^+} + N_n^{b^-} + 4$,

and the result follows.

8.6 Height of the cylinder process when $U(0) \geq \frac{1}{2}$

The results of the preceding section have nothing to say about the height process when $U(0) \geq \lambda^{-1}$. In the next two sections we develop some results for this case. The theorem below is taken from [61].

Theorem 8.19 Let U be a time coordinate distribution with $U(0) \geq \frac{1}{2}$. Then

$\{ \left(\frac{\underline{h}_n^s}{n}\right)^m \}$ is uniformly integrable for all

$m > 0$.

In particular, $E(\frac{\underline{h}_n^s}{n}) \leq \frac{8}{3}$.

Proof. Let n be fixed. Consider subgraphs G_k for $k = 1,2,\ldots$ of the square lattice, where G_k consists of all vertices and bonds in the region bounded by the lines $x = 0$, $x = n$, $y = kn$ and $y = (k-1)n + 1$. Each G_k is isomorphic to the $(n-1) \times n$ sponge defined in §2.1.

When $p = \frac{1}{2}$, Theorem 2.2 implies that the sponge crossing probability $S_{\frac{1}{2}}(n-1,n) = \frac{1}{2}$. Thus if $p = U(0) \geq \frac{1}{2}$, $S_p(n-1,n) \geq \frac{1}{2}$, so with probability at least one-half there is a path in G_k from $x = 0$ to $x = n$ with zero travel time. Such a path serves as a barrier, so there is some route for s_{on} which does not cross it.

Since the subgraphs G_k are disjoint, the events $A_k \equiv \{\exists$ a zero path in G_k from $x = 0$ to $x = n\}$ are independent. Let

$$N^+(\omega) = \inf \{k: \omega \in A_k\}.$$

Then N^+ has a geometric distribution with parameter at least one-half. If H_k is the reflection of G_k in the x-axis, let N^- be defined similarly. Then $\underline{h}_n^s \leq n(\max (N^+,N^-))$ and N^+ and N^- are i.i.d. geometric random variables. Thus all moments of $\frac{h_n^s}{n}$ exist and are uniformly bounded in n, implying the uniform integrability of $\{ \frac{\underline{h}_n^s}{n}^\alpha \}$ for any $\alpha > 0$. A simple calculation shows that $E\{\max (N^+, N^-)\} = \frac{8}{3}$ if $U(0) = \frac{1}{2}$.

8.7 The height process when $U(0) > p_H$

For time coordinate distributions with $U(0) > p_H$, much stronger results than those of the previous sections are possible.

Theorem 8.20 Let U be a time coordinate distribution with $U(0) > p_H$. Then for any $\alpha > 0$,

$$\lim_{n\to\infty} \frac{\underline{h}_n}{n^\alpha} = 0 \quad \text{a.s.}$$

Proof. The proof is given first for the s-process. Let $p = U(0)$, $q = 1-p$, and regard a bond as open if its time coordinate is zero and closed otherwise.

Fix $\beta > 0$. For each positive integer n, let R_n denote the portion of the lattice in $0 \leq x \leq n$, $0 \leq y \leq [n^{\beta}]$, and let R_n^* denote the portion of the dual lattice in $\frac{1}{2} \leq x \leq n-\frac{1}{2}$, $-\frac{1}{2} \leq y \leq [n^{\beta}] + \frac{1}{2}$.

Let n be fixed momentarily. Let $|C|$ denote the number of bonds in the closed cluster containing the origin. From Theorem 3.8, we know that $p > p_H$ implies $q < p_T$, so all moments of $|C|$ are finite. Since there are n vertices in the lower boundary of R_n^*,

$$P\{\text{closed path} \updownarrow R_n^*\} \leq n\, P\{|C| \geq [n^{\beta}] + 1\}$$

$$\leq n \frac{E(|C|^{\frac{3}{\beta}})}{n^3} = \frac{E(|C|^{\frac{3}{\beta}})}{n^2}$$

by Markov's inequality. Summing over n and using Borel-Cantelli,

$$P\{\text{closed path} \updownarrow R_n^* \text{ finitely often}\} = 1.$$

However, if a closed path does not cross R_n^*, an open path must cross R_n from left to right, by Corollary 2.3; in this case there must be a route for s_{on} which does not go above this open path.

An identical argument can be made using a rectangular region below the x-axis, so

$$P\{\underline{h}_n^{s} \geq [n^{\beta}] \text{ finitely often}\} = 1.$$

We conclude that

$$\limsup_n \frac{h_n^s}{n^\beta} \le 1 \quad \text{a.s.} \quad \text{for all } \beta > 0,$$

so that $\lim_n \frac{h_n^s}{n^\alpha} = 0$ a.s. for all $\alpha > 0$. The proof for t_{on} is identical.

For the unrestricted processes, we consider the parts of the lattice L contained in four rectangular regions:

$$-[n^\beta] \le x \le n + [n^\beta], \quad 0 \le y \le [n^\beta]$$

$$-[n^\beta] \le x \le n + [n^\beta], \quad -[n^\beta] \le y \le 0$$

$$-[n^\beta] \le x \le 0 \quad , \quad -[n^\beta] \le y \le [n^\beta]$$

$$n \le x \le n + [n^\beta] \quad , \quad -[n^\beta] \le y \le [n^\beta].$$

By an argument like that above, the probability is one that for n sufficiently large, each of these rectangles is crossed lengthwise by an open path. These open paths link up to form an open circuit enclosing the origin and (n,0), and the enclosed region must contain at least one route for each of a_{on} and b_{on}.

Corollary 8.21 Let U be a time coordinate distribution with $U(0) > p_H$. Then $\lim_{n\to\infty} \frac{h_n}{n} = 0$ in L^p for all $p > 0$.

Proof. It follows from (8.29), (8.32) and their cylinder analogues, from the fact (Lemma 3.12) that $|C|$ has moments of all orders, and from Lemma 8.12, that

$$\left\{E\left(\frac{h_n}{n}\right)^p\right\} \text{ is uniformly bounded in } n, \text{ for any } p > 0.$$

Now apply Theorem 8.20.

8.8 Further results on continuity of $\mu(U)$

As promised at the end of Chapter VII, we will use the work of §8.3 to prove further results about the continuity of $\mu(U)$ as a function of U. As our first application, we extend Theorem 7.12 to the case when U need not be bounded below.

Theorem 8.22 Let U be a distribution with $U(0) < \lambda^{-1}$, and let $\mu(U)$ be the time constant of this distribution. Let U^B be the distribution obtained from U by truncating above at B, i.e.,

$$\begin{aligned} U^B(x) &= U(x) \quad \text{if } x < B \\ &= 1 \quad \text{if } x \geq B \end{aligned}$$

and let $\mu(U^B)$ be the time constant of this distribution. Then

$$\lim_{B\to\infty} \mu(U^B) = \mu(U).$$

Proof. For any $\varepsilon > 0$,

(8.34) $$\frac{a_{on}(\omega)}{n} \le \frac{a_{on}(\omega\oplus\epsilon)}{n} \le \frac{a_{on}(\omega)}{n} + \epsilon \frac{N_n^{a}(\omega)}{n},$$

so that,

(8.35) $$\mu(U) \le \mu(U\oplus\epsilon) \le \mu(U) + \epsilon\, k,$$

with k chosen for the distribution U as in Theorem 8.4. Likewise,

(8.36) $$\mu(U^B) \le \mu(U^B\oplus\epsilon) \le \mu(U^B) + \epsilon k^B.$$

Lemma 8.23 Fix $D > 0$. Then k^B and k can both be chosen not greater than k^D if $B \ge D$.

Proof. Choose k^D in Theorem 8.4 so that $\gamma_D(\frac{\rho}{k^D}) < C^2\lambda^{-1}$, where $C < 1$ and $\rho > \mu(U) \ge \mu(U^D)$. Let X be a random variable with the distribution U, and X^D one with distribution U^D. Then

$$\gamma_D(y) = \inf_{u\ge 0} E(e^{-u(X^D-y)}) = \inf_{u\ge 0} e^{uy} E(e^{-uX^D}),$$

but

$$E(e^{-uX^D}) \ge E(e^{-uX^B}) \ge E(e^{-uX}) \quad \text{if } B \ge D,$$

so that

$$\gamma(y) \ge \gamma_B(y) \ge \gamma_D(y).$$

Therefore, if $\gamma_D(\frac{\rho}{k^D}) < C^2\lambda^{-1}$ it follows that $\gamma_B(\frac{\rho}{k^D}) < C^2\lambda^{-1}$ and $\gamma(\frac{\rho}{k^D}) < C^2\lambda^{-1}$, so k and k^B can both be chosen not greater than k^D.

Returning to the proof of Theorem 8.22, from (8.35), (8.36), and Lemma 8.23 we have

$$\mu(U^B \oplus \epsilon) \le \mu(U^B) + \epsilon\, k^D \le \mu(U) + \epsilon\, k^D$$

$$\le \mu(U \oplus \epsilon) + \epsilon\, k^D \le \mu(U) + 2\epsilon\, k^D$$

Let $B \to \infty$ and apply Theorem 7.12. Then

$$\mu(U) \le \mu(U \oplus \epsilon) \le \lim_B \begin{Bmatrix}\sup\\ \inf\end{Bmatrix} \mu(U^B) + \epsilon\, k^D \le \mu(U) + 2\,\epsilon\, k^D .$$

Since $\epsilon > 0$ is arbitrary it follows that $\lim_B \mu(U^B) = \mu(U)$.

The second application of Theorem 8.8 will show that the time constant of a Bernoulli distribution with atom at zero of size p is a continuous function of p.

> <u>Theorem 8.24</u> Let U_p be a Bernoulli distribution with atom at zero of size p, and let $\mu(p)$ be the time constant of this distribution. Then $p \to \mu(p)$ is continuous on $[0, \lambda^{-1})$.

<u>Proof</u>. We will prove only the left continuity; the other argument is similar. Let $\{U_i\}$ define a time state for the lattice with Bernoulli (p) distributions, where $p < \lambda^{-1}$. Define a time state $\{V_i\}$ as follows:

(8.37) If $U_i = 1$, then $V_i = 1$.

$$\text{If } U_i = 0,\ V_i = \begin{cases} 1 & \text{with probability } \epsilon p^{-1} \\ 0 & \text{with probability } (p-\epsilon)p^{-1}, \end{cases}$$

the determinations being made independently over the bonds. Clearly the distribution of V_i is Bernoulli $(p-\epsilon)$.

Let r be a route for t_{on}, with time state $\{U_i\}$; let t_{on}^{ϵ}

be the corresponding first-passage time with time state $\{V_i\}$. Then

$$t_{on}^{\epsilon} \leq t_{on} + \sum_{\ell_i \epsilon r} 1_{[V_i = 1,\ U_i = 0]}$$

where the sum is taken over all bonds ℓ_i in the route r. Hence

$$E(\frac{t_{on}^{\epsilon}}{n}) \leq E(\frac{t_{on}}{n}) + \epsilon\, E(\frac{N_n^t}{n}),$$

where N_n^t is the number of bonds in the route r. Letting $n \to \infty$ and using Theorem 8.4,

$$\mu(p-\epsilon) \leq \mu(p) + k\,\epsilon$$

where k depends only on p. Hence

(8.38) $$\limsup_{\epsilon \downarrow 0} \mu(p-\epsilon) \leq \mu(p)$$

But $\mu(p-\epsilon) \geq \mu(p)$ by Theorem 7.8, so

(8.39) $$\liminf_{\epsilon \downarrow 0} \mu(p-\epsilon) \geq \mu(p).$$

By (8.38) and (8.39), $\mu(p-\epsilon) \to \mu(p)$ as $\epsilon \downarrow 0$.

Chapter IX

Other percolation models on the square lattice

Up to this point we have considered only the unoriented bond percolation model on the square lattice. In this chapter we will briefly discuss several other percolation models restricting some or all of the bonds to be oriented.

In sections 9.1 - 9.3, we consider a model limited to the first quadrant of the square lattice. The bonds are completely oriented so that from a site (i,j), only the sites $(i+1,j)$ and $(i,j+1)$ may be reached; thus fluid spreads only upward or to the right. Independent identically distributed non-negative random variables, with finite mean, are assigned to the bonds as travel times. Section 9.1 defines and proves convergence of a first-passage process and the associated reach process for this model. In §9.2 the time coordinates are assumed to have an exponential distribution, producing a Poisson growth model studied by Morgan and Welsh (1965); proofs are given verifying two conjectures of Morgan and Welsh. Section 9.3 mentions briefly a completely oriented Bernoulli percolation model considered by Mauldon (1961) and gives some bounds for the critical percolation probability for this model. Finally in §9.4 we treat a partially oriented percolation model on the half-plane in which fluid can spread upward, downward, or to the right, but not to the left.

9.1 The first-passage process and reach process in the completely oriented model.

Let B_n denote the set of lattice points (i,j) in the first quadrant with $i+j=n$. For each n, define v_n as the first-passage time from the origin to B_n over oriented paths which travel only up or to the right.

Theorem 9.1 For any time coordinate distribution U, $\lim_{n\to\infty} \frac{v_n}{n} = \mu(U)$ a.s., where

$$\mu(U) = \inf_n E(\frac{v_n}{n}) = \lim_n E(\frac{v_n}{n}).$$

Proof. First note that there are only 2^n paths from $(0,0)$ to B_n, so that a route of v_n always exists.

For a site P and a positive integer k, define $B'_k(P)$ as the set of lattice points (i',j') in the first quadrant with $i'+j'=k$, where i' and j' are the x and y-coordinates in a translated coordinate system with P as the new origin.

For any pair of positive integers (k,ℓ), let $v'_{k,\ell}$ denote the first-passage time from P_k, the end point of a chosen route for v_k, to $B'_\ell(P_k)$. Since $B'_\ell(P_k) \subset B_{k+\ell}$,

$$v_{k+\ell} \leq v_k + v'_{k,\ell}$$

It is easily seen that v_k and $v'_{k,\ell}$ are independent, that $v'_{k,\ell}$ has the same distribution as v_ℓ, and that v_k increases with k. Also, by considering paths along the x-axis and the y-axis to B_n, an argument very similar to that used in §4.2 will show that $E(v_n^2) < \infty$ for all n. We can therefore apply Theorem 2.9 to show that

$$\lim_{n\to\infty} \frac{v_n}{n} = \mu(U) \quad \text{a.s.},$$

where $\mu(U) = \inf_n E(\frac{v_n}{n}) = \lim_n E(\frac{v_n}{n})$. As before, $\mu(U)$ will be called the time constant of the process.

Define the reach process corresponding to v_n by

$$N_t = \sup \{n\colon v_n \leq t\},$$

representing the most distant boundary B_n that can be reached in time t. (The notation N_t is that used by Morgan and Welsh in [39] and should not be confused with the notation N_n previously used for route length). The treatment of the reach process closely parallels that given in Chapter VI for the unoriented model and the proofs will only be sketched.

<u>Theorem 9.2</u> Let U be any time coordinate distribution. Then

$$\lim_t \frac{N_t}{t} = [\mu(U)]^{-1} \quad \text{a.s.}$$

If $U(0) < \frac{1}{2}$, convergence also takes place in any mean.

Proof. First note that by Theorem 9.1, it can be shown exactly as in Theorem 6.3 that for any time coordinate distribution,

$$\lim_{t\to\infty} \frac{N_t}{t} = [\mu(U)]^{-1} \quad \text{a.s.}$$

Now the connectivity constant for this model is clearly two, since there are 2^n self-avoiding paths of length n; Lemma 6.5 then applies to show that

$$\sup E[\frac{N_t^k}{t^k}] \leq C_k < \infty \quad \text{for} \quad k = 1,2,\ldots$$

and convergence in mean follows as before.

Corollary 9.3 $\mu(U) > 0$ whenever $U(0) < \frac{1}{2}$.

Reh ([40]) has observed that the analogue of Theorem 6.10 also holds for this model, without the restriction that U be bounded above.

Theorem 9.4 Suppose U is a time coordinate distribution with $\mu(U) > 0$. Then

$$\lim_{t\to\infty} E[\frac{N_t}{t}] = [\mu(U)]^{-1}.$$

Proof. If U is bounded above, the proofs of Lemma 6.8, Corollary 6.9, and Theorem 6.10 carry over in an obvious fashion.

If U is not bounded above, truncate U above at B (as in Theorem 7.12) by reducing all time coordinates which are greater than

B to the value B. Let the truncated distribution be denoted by U^B, and the reach process for the new time coordinates by N_t^B. Clearly $N_t^B \geq N_t$ for all t. By Fatou's lemma and the previous case,

$$(9.1) \qquad [\mu(U)]^{-1} \leq \liminf_{t\to\infty} E\left(\frac{N_t}{t}\right) \leq \limsup_{t\to\infty} E\left(\frac{N_t}{t}\right) \leq \limsup_{t\to\infty} E\left(\frac{N_t^B}{t}\right)$$

$$\leq [\mu(U^B)]^{-1}.$$

(Note that $\mu(U^B)$ need not be positive). To complete the proof, we need to show that $\mu(U^B) \to \mu(U)$ as $B \to \infty$. Such a result was obtained in Theorem 7.12 under the condition that U was bounded below away from zero; this condition was needed to get a bound on the route length. In the present model the route length is exactly n, so the condition is unnecessary, and the proof of Theorem 7.12 shows that $\mu(U^B) \to \mu(U)$ as $B \to \infty$ for any time coordinate distribution U. In view of (9.1), Theorem 9.4 is now proved.

Finally we note that the results of this section may be extended to the case where the time coordinates of the horizontal bonds have a different distribution than those of the vertical bonds.

9.2 The Morgan-Welsh model

Morgan and Welsh (1965) considered a special case of the oriented bond problem of the previous section, intended to model the rate of spread of an infection when the spreading mechanism is a Poisson process.

Initially only the origin is infected. The origin infects (0,1) and (1,0) independently according to a Poisson process with parameter λ, i.e.,

$$P((0,1) \text{ infected by time } t) = P((1,0) \text{ infected by time } t) = 1-e^{-\lambda t}.$$

Infection is transmitted only in the direction of the positive x and y-axes, so once infected, (i,j) infects (i+1,j) and (i,j+1) independently in the same manner in which the origin infects (0,1) and (1,0).

It is not difficult to see that the above model corresponds to the completely oriented bond percolation model in the first quadrant, when the time coordinate distributions are exponential with parameter λ.

In the notation of Morgan and Welsh, $M(t,\lambda)$ denotes the mean $E(N_t)$ when the parameter of the exponential distribution is λ. Because λ is simply a scale factor, we can without loss of generality consider $M(t) = M(t,1)$.

Based on Monte Carlo experiments, Morgan and Welsh conjectured that there exists a finite constant C such that

$$\lim_{t\to\infty} \frac{M(t)}{t} = C.$$

This conjecture was verified by Hammersley (1966). An application of Theorem 9.4 and Corollary 9.3 (which is based on Hammersley's method) establishes the conjecture for all distributions with $U(0) < \frac{1}{2}$ and identifies the limit C as the reciprocal of the time constant. Theoretical bounds due to Morgan and Welsh and to Hammersley for the exponential distribution with parameter one yield

$$2.18 \leq C \leq 4.31.$$

If $N_t = k$, the <u>frontier</u> at time t, denoted by r_t, is the number of infected lattice points on B_k. Letting $R(t,\lambda)$ denote $E(r_t)$, Morgan and Welsh showed that $R(t,\lambda)$ exists and conjectured that N_t and r_t are related by

(9.2) $$\frac{dM(t)}{dt} = 2\,R(t).$$

It is clear from §9.1 that $R(t)$ exists: since $r_t \leq N_t + 1$, for any distribution U with $U(0) < \frac{1}{2}$ (or with $\mu(U) > 0$) it follows from Corollary 9.3 (or Theorem 9.4) that $E[r_t] < \infty$ for all t.

The conjecture (9.2) was also proved by Hammersley (1966); we give the proof below.

$$\text{Let}\quad P_n(t) \equiv P(N_t = n)$$

$$P_n{}^k(t) \equiv P(k \text{ sites of } B_n \text{ are infected at time } t, \text{ but } B_{n+1} \text{ is not infected}).$$

It is evident that

(9.3) $$P_n{}^k(t) = 0 \quad \text{for} \quad k > n+1$$

and

(9.4) $$P_n(t) = \sum_{k=1}^{n+1} P_n{}^k(t).$$

Since in a small time interval Δt an infected point can infect at most two other lattice points,

(9.5) $$P_n(t+\Delta t) = \sum_{k=1}^{n+1} P_n^k(t)\,(1-2k\Delta t) + \sum_{k=1}^{n} P_{n-1}^k(t)2k\Delta t + o(\Delta t)$$

Combining (9.3) - (9.5) and letting $\Delta t \to 0$,

(9.6) $$P'_n(t) = -2\sum_{k=1}^{n+1} k\,P_n^k(t) + 2\sum_{k=1}^{n} k\,P_{n-1}^k(t).$$

Let $R_n(t)$ be the conditional expectation of the number of infected points in the boundary, given that the boundary is B_n at time t. Then (9.6) may be rewritten as

(9.7) $$P'_n(t) = -2R_n(t)P_n(t) + 2R_{n-1}(t)P_{n-1}(t).$$

Given that B_n is the boundary at times t and $t+\Delta t$, the probability that the number of infected points in B_n will change during the time interval $(t,t+\Delta t)$ is at most $2n(\Delta t) + o(\Delta t)$. Furthermore, if a change occurs, the magnitude of the change will not exceed n, so

$$R_n(t+\Delta t) - R_n(t) = 0\,(2n^2\Delta t).$$

Thus $R_n(t)$ is a continuous function of t for each fixed n, and by (9.7) $P'_n(t)$ is continuous in t.

By summing (9.7) and telescoping terms in the sum we get

(9.8) $$\sum_{n=0}^{\nu} nP'_n(t) = 2\sum_{n=0}^{\nu-1} R_\nu(t)P_\nu(t) - 2\nu R_\nu(t)P_\nu(t).$$

Next we find an upper bound for $P_n(t)$ for $0 \le t \le T$. Note that the number of self-avoiding walks of k steps from the origin is 2^k. By Theorem 2.11, for the exponential distribution we have

$$\lim_{y \to 0} \gamma(y) = 0.$$

By an application of Lemma 2.12,

$$P_n(t) = P(N_t = n) \leq P(N_t \geq n) = P(v_n \leq t)$$

$$\leq 2^n \gamma(\tfrac{T}{n})^n \quad \text{for} \quad 0 \leq t \leq T.$$

For n sufficiently large, $\gamma(\frac{T}{n}) < \frac{\varepsilon}{2}$, so

$$P_n(t) < \varepsilon^n$$

for some $\varepsilon < 1$.

Since $0 \leq R_n(t) \leq n+1$,

$$2R_\nu(t)P_\nu(t) \leq 2(\nu+1)\varepsilon^\nu \to 0 \quad \text{as} \quad \nu \to \infty$$

uniformly in $t \in [0,T]$. Also,

$$\sum_{n=0}^{\infty} R_n(t)P_n(t) \quad \text{converges uniformly in} \quad 0 \leq t \leq T.$$

Therefore, by (9.8), $\sum_{n=0}^{\infty} nP'_n(t)$ is a uniformly convergent series of continuous functions in $[0,T]$, so

$$M(t) = \sum_{n=0}^{\infty} n\, P_n(t)$$

may be differentiated term by term to get

$$\frac{dM(t)}{dt} = 2 \sum_{n=0}^{\infty} R_n(t)P_n(t) = 2\, R(t) \quad \text{for} \quad t \leq T$$

and in addition, $R(t)$ is continuous. Since T is arbitrary, (9.2) follows.

9.3 Asymmetric oriented percolation

Mauldon (1961) considered an asymmetric completely oriented bond percolation problem on the first quadrant of the square lattice which corresponds to the case of Bernoulli percolation in the model of §9.1. In Mauldon's model the time coordinates of vertical bonds have a different distribution than those of horizontal bonds. The bonds joining lattice points (i,j) to $(i+1,j)$ are open with probability p, while those joining (i,j) to $(i,j+1)$ are open with probability p'. Fluid spreads only upward or to the right, as before, and the origin is taken to be the sole source of fluid.

Mauldon showed that there is a curve passing through the points $(0,1)$ and $(1,0)$ such that if the point (p,p') lies below the curve, with probability one only a finite number of sites are wetted, but if (p,p') lies above the curve, there is a non-zero probability that an infinite set of sites will be wet.

For the symmetric case $(p = p')$, the value of the critical probability (corresponding to p_H in Chapter III) for the completely oriented percolation model is given by the intersection of the critical curve of Mauldon and the line $p = p'$. In this case the lower bound .6198 for the critical probability is attributed by Bishir (1963) to Mauldon, and Hammersley (1959) gave the upper bound .849585.

9.4 A "stiff" percolation model

Consider the square lattice in the half-plane $\{x \geq 0\}$. If fluid is introduced at one or more sites along the y-axis and flows under a directional force applied in the positive x-direction, the corresponding percolation model will have unoriented vertical bonds and horizontal bonds oriented to permit passage only to the right (if we rotate this model clockwise by ninety degrees we would have a model of percolation under the force of gravity). The paths that the fluid travels are then self-avoiding random walks which are "stiff" (i.e., permit no backtracking) in the x-direction. As usual we assign i.i.d., non-negative random variables to the bonds as time coordinates; the processes $\tilde{t}$, $\tilde{s}$, $\tilde{a}$, and $\tilde{b}$ can then be defined for this model exactly as t, s, a, and b are defined in Chapter IV.

The proofs given in Chapter V can be modified to show that the first-passage processes in this case all converge a.s. (when divided by n) to the same constant $\tilde{\mu}(U)$, which depends on the distribution U of the time coordinates. The connectivity constant $\tilde{\lambda}$ for this lattice is clearly bounded above by λ, the connectivity constant for the unoriented lattice. Fisher and Sykes (1959) gave the lower bound

$$\tilde{\lambda} \geq 2.4142.$$

If ν_n represents the number of distinct self-avoiding stiff paths of n steps from $(0,0)$, Kesten (1963) has shown that $\lim_n \frac{\nu_{n+1}}{\nu_n} = \tilde{\lambda}$.

The renewal theory of Chapter VI can be carried through for the stiff model, and route length results corresponding to those of Chapter VIII can be derived when $U(0) < \tilde{\lambda}^{-1}$, as well as height process results like those of §8.5.

The most obvious, and perhaps the most interesting, question for this model concerns the relation between $\mu(U)$ and $\tilde{\mu}(U)$. It is clear from Lemma 1.1 that

$$t_{on}(\omega) \leq \tilde{t}_{on}(\omega) \quad \text{for all } n \text{ and } \omega,$$

hence that

$$(9.9) \qquad \mu(U) \leq \tilde{\mu}(U).$$

It is possible that equality holds in (9.9) for a large class (perhaps all (?)) of distributions U. A proof of this result would have interesting consequences for the route length problem, since it would imply that routes (or near-routes) can be found containing $o(n)$ steps in the negative x-direction in cases when $\mu(U) > 0$.

Chapter X

Conjectures and open problems

It will be evident to any reader of this tract that the theory of first-passage percolation is far from complete. In this chapter we will discuss a few questions which arise more or less naturally from the results of the preceding chapters. (The best-known unsolved problem in the theory is of course the determination of the critical probabilities p_T and p_H discussed in Chapter III.)

10.1 The determination of $\mu(U)$

The results of Chapter VII leave open many questions related to the determination of the time constant $\mu(U)$. It goes without saying that it would be of great interest to determine μ exactly in some non-trivial case, when $\mu > 0$; this appears to be a formidable task. A question that ought to be more tractable is suggested by the results of §7.1:

(10.1) If $\lambda^{-1} \leq U(0) < p_T$, must $\mu(U)$ be strictly positive?

We conjecture that the answer to (10.1) is affirmative. Since by Theorem 7.3, $\mu(U) = 0$ for $U(0) > p_T$, an affirmative answer would imply that p_T is a "critical probability" for first-passage percolation in general.

If it could be shown that $E(\eta_t) = O(t)$ when $U(0) < p_T$, where η_t is defined in §6.2, it would then follow from Fatou's Lemma and Theorem 6.3 that $\mu(U) > 0$. Theorem 6.11 shows that $E(\eta_t) = o(t^{1+\delta})$ for any $\delta > 0$, but this is not quite sufficient.

Examination of Theorem 8.2 shows that $\mu'(0)$ is also a quantity of considerable interest:

(10.2) Are there non-degenerate distributions for which $\mu'(0) = 1$?

From Theorem 8.2, $\mu'(0) = 1$ would imply $\frac{N_n}{n} \to 1$ a.s. if $U(0) < \lambda^{-1}$. We suspect that this cannot occur, hence that the answer to (10.2) is negative.

10.2 The rate of convergence to $\mu(U)$

Theorem 4.1 shows that $\frac{t_{on} - n\mu}{n} \to 0$ a.s. as $n \to \infty$, with similar results for the other first-passage processes. An obvious question is

(10.3) What is the exact rate of convergence of (say) $t_{on} - n\mu$?

In particular, one might conjecture that a central limit theorem holds for the first-passage processes (by Proposition 4.1, this would require a second moment of U):

(10.4) Does $\frac{t_{on} - E(t_{on})}{\sqrt{n}}$ have an asymptotically normal distribution for some class of distributions U?

There is a central limit theorem for subadditive processes, due to Ishitani (1977); however, its application in this context would require knowing that $E(t_{on}) - n\mu$ is $o(n^{\frac{1}{2}})$. The only known example of a central limit theorem for a (non-additive) subadditive process is due to Furstenberg and Kesten (1960), for products of random matrices; using his result Ishitani was able to weaken the conditions imposed on the matrices by Furstenberg and Kesten.

Some doubt on the validity of (10.4) in general was cast by a Monte Carlo study of Welsh (1965), which suggests that $var(t_{on})$ is $o(n)$ for the uniform $(0,1)$ distribution. Hammersley (1974) has noted that some subadditive processes (e.g., the one occuring in Ulam's problem in [35]) appear to have deviations of much smaller order than $O(n^{\frac{1}{2}}\log\log n)$.

10.3 Renewal theory

The results of Chapter VI include weak renewal theorems for the "reach processes" x_t, y_t, x_t^u, and y_t^u. Hammersley and Welsh (1965) raised the question of proving a percolation analogue of the "strong renewal theorem" of Blackwell ([11], Chapter XI):

(10.5) Does $E(x_{t+h}) - E(x_t) \xrightarrow{t\,\infty} h[\mu(U)]^{-1}$ for any $h > 0$?

There seems to be no obvious handle on this problem. The only known strong renewal theorem for subadditive processes is due to Kesten (1973), again in the context of random matrices.

From Theorem 6.7 it follows that if $U(0) < \lambda^{-1}$, $var(x_t)$ is $o(t^2)$ as $t \to \infty$. In classical renewal theory, a much stronger result holds: $var(N_t) \sim Ct$ as $t \to \infty$. Hammersley and Welsh therefore asked:

(10.6) Does $var(x_t) \sim Ct$ for some constant C as $t \to \infty$?

10.4 The constant $\mu(U)$ as a functional of U

A strong result about the continuity of $\mu(U)$ as a functional of U would doubtless be very useful. The following may not be too much to hope for:

(10.7) If distributions U_n converge weakly to U (in the sense of Billingsley (1968)), does $\mu(U_n) \to \mu(U)$?

Even in the special case of Bernoulli percolation, the answer is only partially given by Theorem 8.24:

(10.8) For Bernoulli distributions U_p, is $p \to \mu(p)$ continuous on $[\lambda^{-1}, \frac{1}{2}]$?

It would be surprising indeed if the answer to (10.8) turned out to be negative.

Hammersley and Welsh (1965) raised a question concerning the concavity of the functional $\mu(U)$:

(10.9) if p and q are positive and $p+q=1$, is $\mu(pU_1 + qU_2)$

$$\leq p\mu(U_1) + q\mu(U_2) \ ?$$

They conjectured that the answer to (10.9) is affirmative.

10.5 Route length and the height process

The results of Chapter VIII on route length leave an obvious gap in the case when $\lambda^{-1} \leq U(0) \leq p_H$.

(10.10) Is it true for all distributions U that

$$\limsup_n \frac{N_n(\omega)}{n} \leq k \quad \text{a.s.} \quad \text{for some constant } k \ ?$$

If the answer to (10.10) is yes, the existence and identification of a limit remains to be established:

(10.11) Does Theorem 8.2 extend to (at least some) distributions with $U(0) \geq \lambda^{-1}$?

For the height process, we showed in Chapter VIII that $U(0) > p_H$ implies that $\lim_n \frac{h_n(\omega)}{n^\delta} = 0$ a.s. for any $\delta > 0$.

(10.12) Is it true for all distributions U that $\lim_n \frac{h_n(\omega)}{n} = 0$ a.s.

We suspect that the answer to (10.12) is affirmative. This suspicion arises in part from Theorem 8.2, which shows that the point-to-point and point-to-line processes have, asymptotically, the same number of bonds.

A related "height process", defined below, was introduced by Hammersley and Welsh:

<u>Definition 10.1</u> Let r be a route of $s_{on}(\omega)$.

Let

$$\tilde{h}_n^s(r,\omega) \equiv |m| \text{ if the terminal site of } r$$

is (n,m).

$$\tilde{h}_n^s(\omega) \equiv \inf\{\tilde{h}_n^s(r,\omega): r \text{ is a route of } s_{on}\}.$$

$\tilde{h}_n^b$ is defined similarly.

Hammersley and Welsh pointed out that knowledge of the behavior of $E[\tilde{h}_n^{s,b}]$ as $n \to \infty$ would give information about the rate of convergence of $E(\frac{t_{on}}{n})$ and $E(\frac{a_{on}}{n})$. Of course, an affirmative answer to (10.12) would imply in particular that $\frac{\tilde{h}_n^{s,b}(\omega)}{n} \to 0$ a.s. as $n \to \infty$.

An apparently very difficult problem raised by Hammersley and Welsh is

(10.13) What is the probability $p_n(r)$ that a route of $t_{on}(\omega)$ has exactly r bonds?

It seems that no progress has been made on this problem.

10.6 Richardson's model and the value of $\mu(U)$

Richardson (1974) considered a class of growth models in Euclidean space. One of his models is equivalent to a percolation process on the square lattice with an exponential time coordinate distribution. If fluid is introduced at (0,0) only, one can then examine the configuration of all sites which have been wetted by a certain time t - call it $C(t)$. Richardson proved that there is a norm d on the plane such that, given any $\epsilon > 0$, the probability that the d-ball of radius $(1-\epsilon)t$ is contained in $C(t)$ and $C(t)$ is contained in the d-ball of radius $(1+\epsilon)t$ tends to one as t tends to infinity.

The shape of this limiting configuration is not evident from Richardson's work. Given a site (i,j) with $|i| + |j| = n$, let $w_{(i,j)}$ be the first-passage percolation time from (0,0) to (i,j). It can be shown that, if i and j are chosen so that $\frac{i}{j} \to \theta$ as $n \to \infty$, where $0 \leq \theta \leq \infty$, then

$$\frac{w_{(i,j)}}{n} \to \mu(\theta),$$

where $\mu(\theta)$ is a constant (clearly $\mu(0)$ is the time constant $\mu(U)$ of the exponential distribution). If $\mu(\theta)$ is independent of θ, as in the case of a degenerate distribution, $C(t)$ is a diamond, with vertices along the x and y-axes. If $\mu(\theta) < \mu(0)$ for $0 < \theta < \infty$, as seems to us more likely, the sides of the diamond will bulge outward in the limiting shape. For this and other reasons it would be of interest to know:

(10.14) How does $\mu(\theta)$ vary (if at all) as θ varies between zero and infinity?

In closing, we hopefully suggest to the reader (with apologies to W. S. Gilbert): "Winnow all our folly, folly, folly and you'll find - a grain or two of truth among the chaff."

Appendix A. The FKG inequality

Here we will prove the inequality stated in §2.2, which is due to Fortuin, Kasteleyn, and Ginibre (1971). Since it is no doubt useful for the study of percolation processes on lattices other than the square lattice, we will prove it in its original formulation as a general result about correlations on a finite distributive lattice.

A subset Γ' of a lattice Γ is called a <u>sublattice</u> of Γ if for any x and y in Γ', $x \vee y$ and $x \wedge y$ also lie in Γ'. A subset Γ' of a lattice is called a <u>lower layer</u> (<u>semi-ideal</u> in the terminology of [15]) if whenever $x \in \Gamma'$ and $y \in \Gamma$ with $y \leq x$, then $y \in \Gamma'$. The length of a totally ordered set of n elements is defined as the least upper bound of the lengths of the totally ordered subsets of Γ. If Γ is finite and non-empty, it has a least element 0 and a greatest element I; a minimal element $x \neq 0$ of a lattice is called an <u>atom</u>.

The lemma below is needed in the induction step of the proof of the main result:

> <u>Lemma A.1</u> Let Γ be a finite distributive lattice with an atom a; let Γ'_a, Γ''_a, and Γ_a be the sets $\{x \in \Gamma : x \geq a\}$, $\{x \in \Gamma : x \not\geq a\}$, $\{x \in \Gamma : x = x' \vee a$ for some $x' \in \Gamma''_a\}$, respectively. Then

(i) Γ'_a, Γ''_a, and Γ_a are finite distributive lattices;

(ii) Γ''_a and Γ_a are (lattice) isomorphic; and

(iii) Γ_a is a lower layer of Γ'_a.

Proof (i). It is trivial that Γ'_a is a sublattice of Γ(and hence a lattice). If $x \in \Gamma''_a$ then $x \wedge a < a$ and therefore, since a is an atom, $x \wedge a = 0$; conversely, if $x \wedge a = 0$ then $x \in \Gamma''_a$. If x and y are in Γ''_a then $(x \wedge y) \wedge a = x \wedge (y \wedge a) = 0$; also, $(x \vee y) \wedge a = (x \wedge a) \vee (y \wedge a) = 0 \vee 0 = 0$, so Γ''_a is a sublattice of Γ. Finally, if $x', y' \in \Gamma''_a$, $x = x' \vee a$ and $y = y' \vee a$, then $x \wedge y = (x' \vee a) \wedge (y' \vee a) = (x' \wedge y') \vee a$ and $x \vee y = (x' \vee a) \vee (y' \vee a) = (x' \vee y') \vee a$. So both $x \vee y$ and $x \wedge y$ lie in Γ_a, i.e., Γ_a is a sublattice of Γ. Since Γ is finite and distributive, so are Γ'_a, Γ''_a, and Γ_a.

(ii). If $x \in \Gamma''_a$ then $x \vee a \in \Gamma_a$. Conversely, if $x \in \Gamma_a$ there exists an $x' \in \Gamma''_a$ such that $x = x' \vee a$. Suppose that also $x = x'' \vee a$ with $x'' \in \Gamma''_a$. Then $x' = x' \wedge (x' \vee a) = x' \wedge (x'' \vee a) = (x' \wedge x'') \vee (x' \wedge a) = (x' \wedge x'') \vee 0 = x' \wedge x''$, and by symmetry, $x'' = x' \wedge x''$. Hence $x' = x''$ and it follows that the mapping $x \to x \vee a$ from Γ''_a onto Γ_a is one-to-one; since $x \wedge y \to (x \wedge y) \vee a$ and $x \vee y \to (x \vee y) \vee a$, this mapping is a lattice isomorphism.

(iii). Let $x = x' \vee a$ be an element of Γ_a and suppose that $y \in \Gamma'_a$ with $y \leq x$. Then $(y \wedge x') \vee a = (y \vee a) \wedge (x' \vee a) = y \wedge x = y$;

so $y = y' \vee a$ with $y' = y \wedge x' \leq x'$, and therefore $y' \in \Gamma''_a$. Thus $y \in \Gamma_a$ by definition, and Γ_a is a lower layer of Γ'_a. If I'' is the greatest element of Γ''_a, the greatest element of Γ_a is $I'' \vee a$, and for any $x \in \Gamma_a$ the corresponding element in Γ''_a is $x' = x \wedge I''$.

In the proof of Theorem 2.4 below, no generality is lost in taking μ to be a probability measure on the power set of Γ. If Γ' is a subset of Γ, let

$$E'_\mu(f) \equiv \frac{\sum_{x \in \Gamma'} \mu(x) f(x)}{\sum_{x \in \Gamma'} \mu(x)}$$

<u>Proof of Theorem 2.3</u>. Note first that it is enough that f and g be increasing on the support of μ, i.e., $\{x \in \Gamma: \mu(x) > 0\} \equiv \Gamma_0$, since E_μ is determined only by this set; and (2.3) is trivial unless both x and y are in Γ_0. So we may assume that $\Gamma_0 = \Gamma$, i.e., that μ is strictly positive on Γ.

If Γ has a single element then $\ell(\Gamma) = 0$, and (2.4) holds with equality. Otherwise $\ell(\Gamma) \geq 1$ and Γ contains at least one atom. Assume now that the theorem holds for any lattice of length $\leq n-1$, and let Γ be a lattice of length $n \geq 1$ and μ a strictly positive probability measure on Γ. Let f and g be increasing functions on Γ. Let

(A.1) $\theta = E_\mu(fg) - E_\mu(f)E_\mu(g) = \sum_{x,y \in \Gamma} \mu(x)\mu(y)\,[f(x)g(x) - f(x)g(y)]$.

Let a be an atom of Γ, and let Σ' and Σ'' denote the sum in (A.1) taken over all elements of Γ'_a and Γ''_a, respectively. We can rewrite θ as:

$$\text{(A.2)} \qquad \theta = \sum_x{}' \sum_y{}' \mu(x)\mu(y)\,[f(x)g(x) - f(x)g(y)]$$
$$+ \sum_x{}'' \sum_y{}'' \mu(x)\mu(y)\,[f(x)g(x) - f(x)g(y)]$$
$$+ \sum_x{}' \sum_y{}'' \mu(x)\mu(y)\,[f(x)g(x) - f(x)g(y) + f(y)g(y) - f(y)g(x)].$$

Clearly μ satisfies (2.3) on the sublattices Γ'_a, Γ''_a, and Γ_a, and f and g are increasing on these lattices. Now $\ell(\Gamma'_a) = n-1$ and $\ell(\Gamma''_a) \leq n-1$, so by the induction hypothesis, the first two sums in (A.2) are non-negative. In the third sum we can again use the induction hypothesis on the first and third terms; putting all of this together we have (omitting the summation variables):

$$\text{(A.3)} \qquad \theta \geq \frac{\Sigma'\mu f\, \Sigma'\mu g\, \Sigma''\mu}{\Sigma'\mu} + \frac{\Sigma'\mu\, \Sigma''\mu f\, \Sigma''\mu g}{\Sigma''\mu}$$
$$- \Sigma'\mu f\, \Sigma''\mu g - \Sigma'\mu g\, \Sigma''\mu f,$$

which becomes

$$\text{(A.4)} \qquad \theta \geq (\Sigma'\mu\, \Sigma''\mu)^{-1}\, (\Sigma'\mu f\, \Sigma''\mu - \Sigma'\mu\, \Sigma''\mu f)\, (\Sigma'\mu g\, \Sigma''\mu - \Sigma'\mu\, \Sigma''\mu g).$$

We now claim that all factors on the right-hand side of (A.4) are non-negative, i.e., that

$$\text{(A.5)} \qquad E''_\mu(f) \leq E'_\mu(f), \quad E''_\mu(g) \leq E'_\mu(g).$$

If this can be established it will follow that $\theta \geq 0$, and the theorem will be proved.

To do this, we will show that

(A.6) $$E''_\mu(f) \le E^a_\mu(f) \le E'_\mu(f),$$

where $$E^a_\mu(f) \equiv \frac{\sum_{\Gamma_a} \mu(x)f(x)}{\sum_{\Gamma_a} \mu(x)} .$$

Consider the first inequality in (A.6); if $x \in \Gamma''_a$ and $y \in \Gamma''_a$ with $y \le x$, then (2.3) implies that

(A.7) $$\mu(x)\mu(y\vee a) \le \mu(x\wedge(y\vee a))\mu(x\vee(y\vee a)) = \mu(y)\mu(x\vee a).$$

Therefore, if $\mu_a(x) \equiv \mu(x\vee a)$ for all $x \in \Gamma''_a$, the function $\frac{\mu}{\mu_a}$ is decreasing on Γ''_a. On the other hand, the function $f_a(x) \equiv f(x\vee a)$ is increasing on Γ''_a, since f increases on Γ_a. We apply the induction hypothesis to Γ''_a with the probability measure $\frac{\mu_a}{\Sigma''\mu_a}$, which satisfies (2.3); then

(A.8) $$\Sigma''\mu_a\ \Sigma''\mu f_a \le \Sigma''\mu\ \Sigma''\mu_a f_a .$$

Since f is increasing on Γ, $f \le f_a$ on Γ''_a; this and (A.8) give the first inequality in (A.6).

For the second, observe that

(A.9) $$E^a_\mu(f) = \frac{E'_\mu(f1_{\Gamma_a})}{E'_\mu(1_{\Gamma_a})} .$$

Since Γ_a is a lower layer of Γ'_a, 1_{Γ_a} is decreasing on Γ'_a, so that

$$\text{(A.10)} \qquad E'_{\mu}(f1_{\Gamma_a}) \leq E'_{\mu}(f)\, E'_{\mu}(1_{\Gamma_a}),$$

which by (A.9) gives the second inequality of (A.6).

The authors observe in [15] that the condition in (2.3) is not a necessary condition for increasing functions on Γ to have positive correlations, when $\ell(\Gamma) > 2$.

Appendix B. Ergodic theorems for subadditive processes

We give here a proof of Theorem 2.7 and a sketch of the proof of Theorem 2.8.

> **Lemma B.1** Let $\{x_{mn}\}$ be subadditive with time constant γ, and let $\xi = \lim\sup_n \frac{x_{on}}{n}$. Then ξ is a.s. finite and $E(\xi) = \gamma$; further, $\frac{x_{on}}{n} \to \xi$ in L^1.

Proof. Here we follow Kingman (1968). By subadditivity, for any fixed n,

$$\text{(B.1)} \qquad x_{ot} \leq \sum_{r=1}^{[t/n]} x_{(r-1)n,\, rn} + x_{[t/n]n,t}$$

$$\leq \sum_{r=1}^{[t/n]} x_{(r-1)n,rn} + w_{[t/n]n,t} ,$$

where $w_{[t/n]} = \sum_{u=0}^{n-1} |x_{[t/n]n+u,[t/n]n+u+1}|$.

By the Birkhoff ergodic theorem (using (2.6) and (2.7) of Chapter 2), the limit

$$z_n = \lim_N \frac{1}{N} \sum_{r=1}^{N} x_{(r-1)n,rn}$$

exists a.s., and $E(z_n) = E(x_{on}) = g_n$.

Also, for any $\epsilon > 0$,

$$\sum_{N=1}^{\infty} P(w_N \geq N\epsilon) = \sum_{N=1}^{\infty} P(w_o \geq N\epsilon) \leq \frac{1}{\epsilon} E(w_o) < \infty,$$

so that by the Borel-Cantelli lemma,

$$\frac{W_N}{N} \to 0 \text{ a.s.}$$

Inequality (B.1) then gives

$$\xi \leq \limsup_{t\to\infty} \frac{x_{ot}}{[t/n]n} \leq \limsup_{t\to\infty} \frac{1}{[t/n]n} \{ \sum_{r=1}^{[t/n]} x_{(r-1)n,rn} + w_{[t/n]} \}$$

$$= \frac{z_n}{n}.$$

In particular, $\xi < \infty$ and $E(\xi) \leq \frac{g_n}{n}$. Since this holds for all n,

(B.2) $\quad E(\xi) \leq \gamma$.

Now let

$$b_{st} = \sum_{r=s+1}^{t} x_{r-1,r} - x_{st};$$

b_{st} is then a non-negative superadditive process. If

$$B_n = \inf_{t\geq n} \frac{b_{ot}}{t},$$

then B_n is an increasing sequence, converging as $n \to \infty$ to

$$\liminf_{t\to\infty} \frac{b_{ot}}{t} = \liminf_{t\to\infty} \frac{1}{t} (a_{ot} - x_{ot}) = z_1 - \xi.$$

By monotone convergence,

$$\lim_{n\to\infty} E(B_n) = E(\lim_n B_n) = E(z_1 - \xi) = g_1 - E(\xi).$$

But we also have

$$\lim_{n\to\infty} E(B_n) = \lim_{n\to\infty} E(\frac{b_{on}}{n}) = \lim_{n\to\infty} \frac{ng_1 - g_n}{n} = g_1 - \gamma.$$

Hence $g_1 - E(\xi) \leq g_1 - \gamma$, so that

(B.3) $\quad E(\xi) \geq \gamma.$

From (B.2) and (B.3) we see that $E(\xi) = \gamma$.

Finally,

$$E\left|\frac{b_{ot}}{t} - B_t\right| = E\left(\frac{b_{ot}}{t} - B_t\right) = g_1 - \frac{g_t}{t} - E(B_t)$$

$$\to (g_1 - \gamma) - (g_1 - \gamma) = 0,$$

so that $\frac{b_{ot}}{t} - B_t \to 0$ in L^1 as $t \to \infty$. Since B_t is monotone and converges a.s. to $z_1 - \xi$, it converges in mean, and so $\frac{b_{ot}}{t} \to z_1 - \xi$ in mean. But by Birkhoff's theorem, $\frac{a_{ot}}{t} \to z_1$ in mean, and so

$$\frac{x_{ot}}{t} = \frac{a_{ot} - b_{ot}}{t} \to \xi \quad \text{in } L^1 \text{ as } t \to \infty .$$

For the other half of the proof we follow the argument given by Burkholder in the discussion of [35].

Lemma B.2 There is a stationary sequence $f_o, f_1, \ldots$ such that $E(f_o) = \gamma$ and

$$\sum_{k=s}^{t-1} f_k \leq x_{st} \qquad 0 \leq s < t.$$

Proof. Let

(B.4) $$f_{kn} = \frac{1}{n}\sum_{r=1}^{n} (x_{k,k+r} - x_{k+1,k+r}) .$$

By (2.6), the sequence $\underset{\sim}{f}_o = (f_{on})$, $\underset{\sim}{f}_1 = (f_{1n})$, ..., is stationary. Take t with $t \geq k+1 \geq s+1$, and $n > t$. Since $x_{kr} - x_{k+1,r} \leq x_{k,k+1}$ for all $k+1 \leq r$,

$$nf_{kn} = \sum_{r=k+1}^{k+n} (x_{kr} - x_{k+1,r}) \leq \sum_{r=t+1}^{r} (x_{kr} - x_{k+1,r}) + t\, x_{k,k+1}.$$

Therefore,

$$n \sum_{k=s}^{t-1} f_{kn} \leq \sum_{r=t+1}^{n} (x_{sr} - x_{tr}) + t \sum_{k=s}^{t-1} x_{k,k+1} .$$

By (2.5), the first sum on the right-hand side is dominated by

$\sum_{r=t+1}^{n} x_{st} - (n-t)x_{st}$. Thus

$$\text{(B.5)} \qquad \sum_{k=s}^{t-1} f_{kn} \leq x_{st} + \frac{1}{n} w_{st} \quad \text{provided that } n > t,$$

where $w_{st} = t\,[\sum_{k=s}^{t-1} x_{k,k+1} - x_{st}]$.

Now $f_{on} \leq x_{o1}$ and by (B.4),

$$E(f_{on}) = \frac{1}{n} \sum_{r=1}^{n} (g_r - g_{r-1}) = \frac{1}{n} g_n \geq \gamma .$$

So $E|f_{on}| \leq E|x_{o1}| + E(x_{o1} - f_{on}) \leq E|x_{o1}| + g_1 - \gamma$

and the sequence (f_{on}) is L^1-bounded. By a theorem of Komlòs (1967) there is a sequence $n_1 < n_2 < \ldots$ of positive integers and an integrable function f_o such that

$$A_o^j = \frac{1}{j} \sum_{i=1}^{j} f_{on_i} \to f_o \text{ a.s. as } j \to \infty .$$

By stationarity,

$$A_k^j = \frac{1}{j} \sum_{i=1}^{j} f_{kn_i} \to f_k \text{ a.s. as } j \to \infty$$

and $(f_o, f_1, \ldots)$ is a stationary sequence. Given $t \in N$, let

$$i_t = \inf\{i : t < n_i\}$$

and define

$$A_o^j(t) = \frac{1}{j}\sum_{i=i_t}^{i_t+j-1} f_{on_i}$$

with a similar definition of $A_k^j(t)$. Clearly $A_k^j(t) \to f_k$ a.s. as $j \to \infty$ .for any t and $k \in N$. By (B.5),

$$\sum_{k=s}^{t-1} A_k^j(t) \leq x_{st} + w_{st}\left[\frac{1}{j}\sum_{i_t}^{i_t+j-1} \frac{1}{n_i}\right].$$

Letting $j \to \infty$, we get $\sum_{k=s}^{t-1} f_k \leq x_{st}$.

Let $\{x_{mn}\}$ be subadditive with time constant γ; putting Lemmas B.1 and B.2 together we can prove that $\frac{x_{on}}{n}$ converges a.s. Let $y_{mn} = \sum_{k=m}^{n-1} f_k$, as in Lemma B.2. Then $x_{mn} = y_{mn} + z_{mn}$, where $\{z_{mn}\}$ is non-negative and subadditive. Also,

$E(\frac{z_{on}}{n}) = E(\frac{x_{on}}{n}) - E(\frac{y_{on}}{n}) = \frac{g_n}{n} - \gamma \to 0$ as $n \to \infty$, so the process $\{z_{mn}\}$ has time constant zero.

Applying Lemma B.1, if $\eta = \limsup_{n\to\infty} \frac{z_{on}}{n}$, we have $\eta \geq 0$, since $z_{on} \geq 0$, but $E(\eta) = 0$. Hence $\eta = 0$ a.s. and $\frac{z_{on}}{n} \to 0$ a.s. By the ergodic theorem, $\frac{\ \ _{on}}{n} \to \xi$ a.s. and in L^1, so that $\frac{x_{on}}{n} \to \xi$ a.s. and in L^1.

Finally suppose that the process $\{x_{mn}\}$ is independent. Let $\mathcal{T}$ be the σ-field of events generated by $\{x_{mn}\}$ which are invariant under the shift θ: $x_{mn} \to x_{m+1,n+1}$. It is easy to see that ξ is measurable in $\mathcal{T}$; since $\{x_{mn}\}$ is independent, $\mathcal{T}$ is trivial, and it follows that $\xi = \gamma$ a.s. This completes the proof of Theorem 2.7.

The proof of Theorem 2.8 rests on Theorem B.3 below. As observed by Dunford (1951), this theorem is essentially a consequence of a mean ergodic theorem of F. Riesz (1938); the proof is implicit in the proof of the main theorem in Dunford's paper.

<u>Theorem B.3</u> Let $\{x_{\underset{\sim}{j}}\}_{\underset{\sim}{j} \in N^2}$ be an array of random variables stationary under the shifts θ_1: $x_{i,j} \to x_{i+1,j}$ and θ_2: $x_{i,j} \to x_{i,j+1}$. Let $S_{\underset{\sim}{n}} = \sum_{\underset{\sim}{j} \leq \underset{\sim}{n}} x_{\underset{\sim}{j}}$ and, if $\underset{\sim}{n} = (n_1, n_2)$, let $|\underset{\sim}{n}| = n_1 n_2$. Then if $x_{(1,1)}$ is integrable,

$$\frac{S_{\underset{\sim}{n}}}{|\underset{\sim}{n}|} \text{ converges in } L^1.$$

The proof of Theorem 2.8 proceeds as in Lemma B.1, using Theorem B.3 in place of the Birkhoff theorem, and a standard diagonal argument; the details may be found in [47].

REFERENCES

[1] Billingsley, Patrick (1968). Convergence of Probability Measures. Wiley, New York.

[2] Bishir, J. (1963). A lower bound for the critical probability in the one-quadrant oriented-atom percolation process. J. Roy. Stat. Soc. Ser. B 25, 401-404.

[3] Broadbent, S. R. and Hammersley, J. M. (1957). Percolation processes. I. Crystals and mazes. Proc. Camb. Phil. Soc. 53, 629-641.

[4] Chung, K. L. (1974). A Course in Probability Theory. Second edition. Academic Press, New York.

[5] Dean, P. and Bird, N. F. (1967). Monte Carlo estimates of critical percolation probabilities. Proc. Camb. Phil. Soc. 63, 477-479.

[6] Del Junco, A. (1977). On the decomposition of a subadditive stochastic process. Ann. Prob. 5, 298-302.

[7] Derrienic, Y. (1975). Sur le théorème ergodique sous-additif. Comptes Rendus Acad. Sci. Paris Ser. A 281, 985-988.

[8] Dunford, N. (1951). An individual ergodic theorem for non-commutative transformations. Acta Sci. Math. Szeged. 14, 1-4.

[9] Essam, J. W. (1972). Percolation and cluster size. Phase Transitions and Critical Phenomena, Volume 2, C. Domb and M. S. Green, editors, Academic Press, New York.

[10] Essam, J. W. and Fisher, M. E. (1961). Some cluster size and percolation problems. J. Math. Phys. 2, 609-619.

[11] Feller, W. (1971). An Introduction to Probability Theory and its Applications, Volume 2, second edition, Wiley, New York.

[12] Fisher, M. E. (1961). Critical probabilities for cluster size and percolation problems. J. Math. Phys. 2, 620-627.

[13] Fisher, M. and Sykes, M. F. (1959). Excluded volume problem and the Ising model of ferromagnetism. Phys. Rev. (Ser. 2) 114, 45-58.

[14] Fortuin, C. M. (1972). On the random cluster model II. The percolation model. Physica 58, 393-418.

[15] Fortuin, C. M., Kasteleyn, P. W., and Ginibre, J. (1971). Correlation inequalities on some partially ordered sets. Comm. Math. Phys. 22, 89-103.

[16] Frisch, H. L. and Hammersley, J. M. (1963). Percolation processes and related topics. J. SIAM 11, 894-918.

[17] Furstenberg, H. and Kesten, H. (1960). Products of random matrices. Ann. Math. Stat. 31, 457-469.

[18] Grimmett, G. R. (1976). On the number of clusters in the percolation model. J. London Math. Soc. Ser. 2, 13, 346-350.

[19] Hammersley, J. M. (1957). Percolation processes. II. The connective constant. Proc. Camb. Phil. Soc. 53, 642-645.

[20] Hammersley, J. M. (1959). Bornes superieures de la probabilité critique dans un processus de filtration. Le Calcul des Probabilitiés et ses Applications. Centre national de la recherche scientifique, Paris, 17-37.

[21] Hammersley, J. M. (1961A). The number of polygons on a lattice. Proc. Camb. Phil. Soc. 57, 516-523.

[22] Hammersley, J. M. (1961B). On the rate of convergence to the connective constant of the hypercubical lattice. Quart. J. Math. Oxford Ser. 2, 12, 250-256.

[23] Hammersley, J. M. (1963). Long-chain polymers and self-avoiding random walks. Sankhya 25, 15-34.

[24] Hammersley, J. M. (1966). First passage percolation. J. Roy. Stat. Soc. Ser. B 28, 491-496.

[25] Hammersley, J. M. (1974). Postulates for subadditive processes. Ann. Prob. 2, 652-680.

[26] Hammersley, J. M. and Welsh, D. J. A. (1962). Further results on the rate of convergence to the connective constant of the hypercubical lattice. Quart. J. Math. Oxford Ser. 2, 13, 108-110

[27] Hammersley, J. M. and Welsh, D. J. A. (1965). First passage percolation, subadditive processes, stochastic networks, and generalized renewal theory. Bernoulli-Bayes-Laplace Anniversary Volume, J. Neyman and L. M. LeCam, editors, Springer-Verlag, Berlin.

[28] Harris, T. E. (1960). A lower bound for the critical probability in a certain percolation process. Proc. Camb. Phil. Soc. 56, 13-20.

[29] Hille, E. and Phillips, R. S. (1957). Functional Analysis and Semi-Groups, American Mathematical Society, Providence, R. I.

[30] Ishitani, Hiroshi (1977). A central limit theorem for the subadditive process and its application to products of random matrices. Publ. R. I. M. S., Kyoto Univ., volume 12, no. 3, 565-575.

[31] Kesten, Harry (1963). On the number of self-avoiding walks. J. Math. Phys. 4, 960-969.

[32] Kesten, Harry (1964). On the number of self-avoiding walks. II. J. Math. Phys. 5, 1128-1137.

[33] Kesten, Harry (1973). Random difference equations and renewal theory for products of random matrices. Acta Math. 131, 107-148.

[34] Kingman, J. F. C. (1968). The ergodic theory of subadditive stochastic processes. J. Roy. Stat. Soc. Ser. B 30, 499-510.

[35] Kingman, J. F. C. (1973). Subadditive ergodic theory. Ann. Prob. 1, 883-909.

[36] Komlos, J. (1967). A generalization of a problem of Steinhaus. Acta Math. Acad. Sci. Hungar. 18, 217-229.

[37] Kurkijarvi, J. and Padmore, T. C. (1975). Percolation in two-dimensional lattices, J. Phys. A 8, 683-696.

[38] Mauldon, J. G. (1961). Asymmetric oriented percolation on a plane. Proceedings, Fourth Berkeley Symposium, Vol. 2, 337-345.

[39] Morgan, R. W. and Welsh, D. J. A. (1965). A two-dimensional Poisson growth process. J. Roy. Stat. Soc. Ser. B 27, 497-504.

[40] Reh, Wolfgang (1977). Personal communication.

[41] Richardson, Daniel (1974). Random growth in a tessellation. Proc. Camb. Phil. Soc. 70, 515-528.

[42] Riesz, F. (1938). Some mean ergodic theorems. J. London Math. Soc. 13, 274-278.

[43] Saks, Stanislaw (1964). Theory of the Integral, second edition, Dover, New York.

[44] Seymour, P. D. and Welsh, D. J. A. (1977). Percolation probabilities on the square lattice. Proceedings, Cambridge Conference on Combinatorial Theory 1977, to appear.

[45] Shante, V. K. S. and Kirkpatrick, S. (1971). An introduction to percolation theory. Adv. Phys. 20, 325-357.

[46] Smythe, R. T. (1973). Strong laws of large numbers for r-dimensional arrays of random variables. Ann. Prob. 1, 164-170.

[47] Smythe, R. T. (1976A). Multiparameter subadditive processes. Ann. Prob. 4, 772-782.

[48] Smythe, R. T. (1976B). Remarks on renewal theory for percolation processes. J. Appl. Prob. 13, 290-300.

[49] Smythe, R. T. and Wierman, John C. (1977). First passage percolation on the square lattice. I. Adv. Appl. Prob. 9, 38-54.

[50] Smythe, R. T. and Wierman, John C. (1978). First passage percolation on the square lattice. III. Adv. Appl. Prob. 10 (to appear).

[51] Sykes, M. F. and Essam, J. W. (1964). Exact critical percolation probabilities for site and bond problems in two dimensions. J. Math. Phys. 5, 1117-1127.

[52] Sykes, M. F., Gaunt, D. S., and Glen, M. (1976A). Percolation processes in two dimensions. II. Critical concentrations and the mean size index. J. Phys. A 9, 97-103.

[53] Sykes, M. F., Gaunt, D. S., and Glen, M. (1976B). Percolation processes in two dimensions. III. High-density series expansions. J. Phys. A 9, 715-724.

[54] Sykes, M. F., Gaunt, D. S., and Glen, M. (1976C). Percolation processes in two-dimensions. IV. Percolation probability. J. Phys. A 9, 725-730.

[55] Temperley, H.N.V. and Lieb, E.H.(1971). Relations between the 'percolation' and 'colouring' problem and other graph-theoretical problems associated with regular planar lattices: some exact results for the 'percolation' problem. Proc. Roy. Soc. London A 322, 251-280.

[56] Welsh, D.J.A.(1965). An upper bound for a percolation constant. J. Appl. Math. Phys. 16, 520-522.

[57] Welsh, D.J.A.(1977). Percolation and related topics. Science Progress 64, 65-83.

[58] Whitney, Hassler (1932). Nonseparable and planar graphs. Trans. Amer. Math. Soc. 34, 339-362.

[59] Whitney, Hassler (1933). Planar graphs. Fund. Math. 21, 73-84.

[60] Wierman, John C. (1977). First passage percolation on the square lattice. II. Adv. Appl. Prob. 9, 283-295.

[61] Wierman, John C. and Reh, Wolfgang (1978). On conjectures in first passage percolation theory. Ann. Prob. 6. to appear.

Subject Index